SpringerBriefs in Environment, Security, Development and Peace

Volume 23

Series editor

Hans Günter Brauch, Mosbach, Germany

More information about this series at http://www.springer.com/series/10357
http://www.afes-press-books.de/html/SpringerBriefs_ESDP.htm
http://www.afes-press-books.de/html/SpringerBriefs_ESDP_23.htm

Rosario H. Pérez-Espejo
Roberto M. Constantino-Toto
Hilda R. Dávila-Ibáñez
Editors

Water, Food and Welfare

Water Footprint as a Complementary
Approach to Water Management in Mexico

Editors
Rosario H. Pérez-Espejo
Institute of Economics Research
Mexico, Distrito Federal
Mexico

Roberto M. Constantino-Toto
Unidad Xochimilco
Universidad Autónoma Metropolitana
Mexico, Distrito Federal
Mexico

Hilda R. Dávila-Ibáñez
Unidad Xochimilco
Universidad Autónoma Metropolitana
Mexico, Distrito Federal
Mexico

ISSN 2193-3162   ISSN 2193-3170   (electronic)
SpringerBriefs in Environment, Security, Development and Peace
ISBN 978-3-319-28822-2   ISBN 978-3-319-28824-6   (eBook)
DOI 10.1007/978-3-319-28824-6

Library of Congress Control Number: 2015960767

Cover photo: River Lerma in La Piedad, Michoacan, Mexico. The editors hold the copyright
for this photo and granted the permission to use it here. More on this book is at: ⟨http://
www.afes-press-books.de/html/SpringerBriefs_ESDP_23.htm⟩

Copyediting: PD Dr. Hans Günter Brauch, AFES-PRESS e.V., Mosbach, Germany

Printed on acid-free paper

This Springer imprint is published by SpringerNature
The registered company is Springer International Publishing AG Switzerland

Acknowledgements

This book is the result of an invitation in 2012 from the Water Observatory of the Botin Foundation, a private organization in Spain, to the Scientific Water Network of the National Council for Science and Technology, a public institution in Mexico, to participate in a research project regarding water and food security in Latin America. We would like to thank these institutions for making it possible for 19 researchers of 11 Mexican higher education and research institutions to work in an original project concerning water issues.

M. Ramón Llamas from Botín Foundation, as well as Alberto Garrido and Bárbara Willarts, his collaborators, from the Politechnical University of Madrid deserve special mention. Special acknowledgments likewise to Tomás Viveros, Director at that time of the Scientific Networks of the National Council for Science and Technology; Patricia E. Alfaro Moctezuma, Chancellor of Metropolitan Autonomous University, Xochimilco Campus (UAM-X); and Verónica O. Villarespe Reyes, Director of the Institute for Economics Research, at the National Autonomous University of Mexico (UNAM). Also, we want to thank Úrsula Oswald Coordinator of the Scientific Water Network of the National Council for Science and Technology.

We would like to especially acknowledge Hans Gunter Brauch, from the Free University of Berlin and head editor of Hexagon Book Series; Johanna Schwartz, head editor of Springer-Verlag; and Divya Selvaraj, Project Coordinator (books) at Springer, for their support and collaboration, which were essential to the publishing of this book.

This publication would not have been possible without the support of the Editing Department at IIEc, especially Graciela Reynoso Rivas from the Publishing Department of Social Sciences and Humanities Division at UAM-X, particularly Miguel Ángel Hinojosa Carranza; Isaac Alí Siles Barcena for his professional translation; and Citlalin Martínez Córdova for providing us with technical support.

Not only do we thank the researchers who contributed with their expertise throughout the different chapters in the book, but we also apologize to everyone for the long time it took for your documents to be published, first as an e-book in Spanish, and secondly, as an English publication by Springer.

Mexico City, UAM and UNAM   Rosario H. Pérez-Espejo
June 2015   Roberto M. Constantino-Toto
                                                           Hilda R. Dávila-Ibáñez

Contents

Introduction and Structure of the Book

The book *Water, Food and Welfare. Water footprint as a complementary approach to water management in Mexico*, is the revised and updated version of the report issued in November 2012 (the Mexico Report) to the Water Observatory of the *Botín* Foundation (Spanish acronym: OAFB), whose initiative convened seven countries in Latin America, Mexico among them, to participate in the preparation of the book whose final title was *Water for Food Security and Well-being in Latin America and the Caribbean. Social and Environmental Implications for a Globalized Economy*.

The Mexico Report was one of the tasks of the *Water Thematic Network of the National Council for Science and Technology* (Spanish acronym: RETAC), who served as the Mexican counterpart of OAFB and whose coordination was provided by Rosario H. Pérez Espejo. The agreement with OAFB was that they would use the information in the Mexico Report they required for the development of their own book, and we would be able to publish this report on the terms we decided.

This is how most of the 13 collaborators from 10 research and higher education institutions participating in the Mexico Report, made their contributions, in some cases corrected and adjusted for this book to which new researchers joined. The structure, guidance and opinions contained in each chapter are the sole responsibility of its authors and is completely independent of the approaches to Mexico that can be found in the book by the Water Observatory of the *Botín* Foundation.

Editing and publication of this book is the result of collaboration between the Department of Economic Production of the *Metropolitan Autonomous University* (UAM)—Xochimilco, and the Institute of Economic Research of the *National Autonomous University of Mexico* (UNAM).

Mexico City, UAM and UNAM
May 2015

Rosario H. Pérez-Espejo
Roberto M. Constantino-Toto
Hilda R. Dávila-Ibáñez

Structure of the Book

Part I: Linking Water Management, Food Policy and Welfare

Chapter 1 on "Contemporary Model for Water Management and Alternative Approaches" raises the need to complement the capabilities of water management in Mexico by incorporating alternative analytical approaches such as water footprint, water *colors* and virtual water, so as to strengthen the processes for identifying the best practices in the use of this resource, according to potential sources of supply and facilitate setting up new arrangements for the development of public investment and infrastructure required.

In Chapter 2 on "Socioeconomic Framework" an overview of the country is presented from its main economic indicators, highlighting the unequal distribution of income, rural bias of poverty, and the particular importance of the agricultural sector as a user of water resources.

Chapter 3 on "Water Policy and Institutions" displays the Mexican legislation on the use and exploitation of Mexican water resources at different tiers and bodies of government and outlines water policy on the basis of its main instrument, the National Water Plan.

Chapter 4 on "Water use For Food Purposes" considers how the agriculture and cattle raising sector uses water for food in Mexico. It springs from a review of the geographical, social, and political conditions from which farms and ranches in Mexico are producing, emphasizing the heterogeneous composition of the sector and the presence of an external water footprint due to the import of agricultural products

In Chapter 5 on "Water Resources Inventory and Implications of Irrigation Modernization," it is observed that the induction of a policy toward irrigation modernization has an impact on the expansion of irrigation coverage with negative effects on the country's water reserves.

Chapter 6 on "Manifestations of Welfare Loss" shows that food security is oriented toward people and peace, where participatory governance and peaceful conflict negotiation boost the recovery and protection of ecosystems, and science provides methodologies, standards, and laws capable of protecting the survival of humanity.

Chapter 7 on "Prices and Water: A strategy with Limited Effectiveness" studies the characteristics of water markets within the analytical framework of water and alimentary security. Institutional evolution marked by the lack of coordination between different levels of management related to water policy and the phenomenon of financial sustainability required for the promotion of water and food security is analyzed.

Part II: Pressures on Water Availability, Its Use, and Welfare Effects

This part presents a set of processes and factors that exert strong pressure on water use and availability, and consequently in food safety. In Chapter 8, "Water Use Pattern," competition for water between the growing and dynamic nonagricultural uses and the agricultural sector is analyzed.

Chapter 9 on "Change in the Dietary Pattern and Water Security" links three dimensions: food sovereignty, water security, and change in the pattern of food consumption, both theoretically and empirically. It presents a theoretical discussion of the relationship of these dimensions, placing the concept of food security in the center of the argument; it also presents an estimation exercise for the effect of changes in food consumption on food sovereignty and water security in Mexico in 1992 and 2010. The results point to the importance of assessing all dimensions involved in the realization of the right to food.

Chapter 10 on "Hydrological Stress and Pressures on water Availability" considers that water as a scarce resource and public good is a social, political, and economic problem. Its purpose is to illustrate the conditions of water availability and the intended use. It lays a foundation for the analysis of pressures on water resources in order to understand the circumstances that determine a disparity in water availability. Special attention is given to consumptive uses of water, as they put the biggest pressure on this resource. Furthermore, it raises the question of how much water the market requires to satisfy demand, then details some conflicts over water rights at the national and international level. Because of the wide extent of the country, the diverse ecosystems behave differently in terms of water availability and given that the population distribution is unequal, water demand is also uneven.

Chapter 11 on "Problems Associated to Groundwater Management" describes how this resource is managed, emphasizing that although underground water reserves are large, a global count does not reflect the plight of vast arid and semiarid regions, where water balance is negative and underground storage is low. Groundwater concession is analyzed, highlighting the great weight of the volume used in agriculture and an index expressing the degree of pressure from agricultural use of groundwater, as the percentage of groundwater used in other consumptive uses compared to that for agricultural use, is presented. Nationally, this degree of pressure is close to very high, indicating that food security depending on agricultural irrigation with groundwater is subject to strong competition between users. The increased pressure on food security is exercised by the use of Public Water Supply. It concludes that the high degree of pressure on groundwater requires effective management of this resource, which includes actions to increase water availability in aquifers, promote their preservation, comprehensive utilization, use, efficiency, and reuse.

Finally, Chapter 12 on "Vulnerability and Climate Change" analyzes linkages and implications of the potential presence of large-scale hydro-meteorological events in Mexico and the relevance of the water footprint approach to strengthen the institutional capacities of management that tend to mitigate impacts on welfare in local areas. The research addresses the notions of vulnerability and resilience to impacts that alter the stability of the socioeconomic systems. It sets the methodological features that a water footprint approach could bring to strengthen institutional capacities for managing risks from hydro-meteorological impacts, based on cluster analysis, it draws an approximation to latent vulnerability to droughts in the states of the federation.

Part III: Methodology for Analyzing Water Footprint and Virtual Water

This part consists of four chapters proposing, exploring, and applying various methodological aspects of water footprint. Chapter 13 on "Water Demand of Major Crops: A methodology" proposes an algorithm for calculating water requirements of major crops under irrigation ("blue water footprint"), according to local climate and soil moisture balance. It defines and organizes the variables to consider (computer program "DRIEGO 1") and presents in graphical and tabular form, the drought index and evapotranspiration.

Chapter 14 on "Gray Water Footprint and Water Pollution" analyzes water pollution in terms of their use, the interrelation of contaminants to the water environment, and regulatory aspects of pollution, both in water and in its environment. It discusses the concept of "gray water footprint" (Hoekstra and Chapagain, 2008) and determines in what situations it is possible to apply this tool to better understand the impact of water pollution and get useful values for decision-making in a hydrological basin.

Water footprint, defined as the amount of water used directly and indirectly by the mining of precious metals is discussed in Chapter 15 on "Gray Footprint and Mining: Impact of Metal Extraction on Water." The lack of basic information on this sector makes it impossible to estimate the amount of water used, and the effects of multiple pollutants generated in geomorphology and hydrology of a given basin, as well as runoff and water quality.

Finally, Chapter 16 on "Considerations on Virtual Water and Agri-food Trade" argues that the concepts of virtual water and water footprint allow for an instrument that shows the flows of water (real but hidden) that occur in the international trade and consumption of food. However, their uncritical use can justify an ideology of free trade based on comparative advantage in the provision of water resources and the apparent protection of ecosystems, without considering aspects of food security, development and welfare of communities, as well as equity among members of a society.

Part IV: Applying the WF Approach for Impact Analysis on Sectors and Regions

This part implements a key feature of the water footprint (WF) methodology, namely its applicability in specific geographic areas and economic sectors. Chapter 17 on "The Water Footprint of Four Cereals in Irrigation District 011" analyzes the WF of corn, wheat, sorghum, and barley grains occupying 90 % of the surface and a similar proportion of water in one of the major irrigation districts in the country. It utilizes the DRiego Program and contrasts its results with CROPWAT (FAO) and Hoekstra. It is shown that WF estimates are very sensitive to production output and that a low WF does not indicate the actual water use. The lack of information at disaggregated levels limits WF estimation in the agricultural sector.

Chapter 18 on "Forage Water Footprint in the Comarca Lagunera" reveals the irrationality of the fodder-milk system, highly demanding of water, in a semiarid country, where in addition, urban and industrial demand for the resource is growing. The DRiego Program is used to estimate water requirements of fodder for the production of a liter of milk, and energy demand (subsidized) for the production of forage with water from an overexploited aquifer, is estimated.

Chapter 19 on "Water Footprint of Livestock" highlights the dynamic consumption of animal origin produce and estimates the blue and green WF of beef, pork, poultry, milk, and eggs, both domestic and imported. The results are compared with those of Hoekstra. The lack of specific information on water consumption by different farms in different ecological zones and the various production processes limit the estimate of WF, a tool that can be very useful.

The analysis of "Water Footprint of Bottled Drinks and Beverages and Alimentary Security" in Chapter 20 discusses the incongruity between the dynamism of this sector and shortages, quality, and water management in Mexico. The study questions the importance of the bottling sector and of the pattern of consumption of its products, and the virtual water content of the products of this industry is estimated based on Hoekstra (2010) and Garrido (2010), in order to make such approaches comparable in their magnitudes.

Part I
Linking Water Management, Food Policy and Welfare

Chapter 1
Contemporary Model for Water Management and Alternative Approaches

Roberto M. Constantino-Toto

Abstract Water is a natural good with the characteristic of being interpreted in different forms due to its transversality (Some examples of the importance of water based on the multiple ways in which it can be interpreted are, among others: an ecological way, establishing it as an essential factor for the sustaining of life; an economic one, through the exegesis of considering it an essential input in the provision of material goods which constitute the basis of social welfare; an institutional one, interpreted as an element essential to the formation of collective prosperity; a social one, as a vehicle for stability which allows the cohesion and reproduction of the system; a sanitary one, regarding it as a determining factor in the quality of life of the population). Of course, they are all equally important and occur simultaneously, which in turn highlights how significant and essential its availability is. Water availability is at the center of public welfare formation and prosperity evolution in all society. Nevertheless, it is a topic of growing public and institutional interest in the case of arid and semiarid societies, and those with heterogeneity in its distribution, such as the Mexican one. This is because the allocation processes to be encouraged should yield the maximum effect of public welfare from its exploitation process. This chapter puts forth the need to supplement water management capacities in Mexico through the incorporation of alternative analytical efforts such as those of water footprint and virtual water, which may strengthen the processes to identify the best practices in the exploitation of this resource, according to the potential supply sources and may foster the configuration of new arrangements for the required public infrastructure and investments development that aims at reducing vulnerability in water availability, derived from the institutional strategy that consolidated over the long term.

Keywords Water management · Alternative approaches

© The Author(s) 2016
R.H. Pérez-Espejo et al. (eds.), *Water, Food and Welfare*,
SpringerBriefs in Environment, Security, Development and Peace 23,
DOI 10.1007/978-3-319-28824-6_1

1.1 An Institutional Perspective on Water Availability and Its Scarcity

The process of water supply for a society is related, at a first stage, to the natural physical availability of hydric resources. However, not all the water quantified form natural availability is susceptible of being used in the supply. Evapotranspiration, underground filtration, accumulation in superficial bodies, as well as the capacity of the infrastructure actually in place to extract, condition, distribute, and eventually collect water must be deducted from the estimated resource flow from the hydrologic cycle; that is to say, effective water availability in a society is remarkably affected by the technical choices made for its provision.

From an institutional perspective, the water supply process has a first limit in the capacity of the infrastructure for its provision, and therefore, of the underlying costs structure for the system operation, the maintenance of infrastructure, the increase of capacity, and the eventual creation of new facilities and equipment. In that regard, in a scenario of gradual increase of the natural demand of flows resulting from the demographic and productive dynamics, water availability may face an institutional situation of scarcity if the capacity growth rate or the required investment to that end, does not evolve at the same pace as exploitation.

An additional condition that determines supply sufficiency is one related to exploitation practices. Water availability is affected by the utilization pattern among the different types of uses. This condition is the result of the setting of incentives, whether they are explicit or not, that encourage extensive technological patterns of exploitation in the case of water as an input for production activity; or the result of careless consumption practices in the case of domestic uses. Two of the evident results of completion for the use of water resources in the face of a water offer with a steady technological trajectory and without the presence of corrective resources on exploitation are, on the one hand, overexploitation of water reserves: On the other hand, the presence of critical conflicts among users and between them and the authority responsible for the supply systems (Sainz and Becerra 2003).

The availability scenario is completed by the incorporation of the exploitation of waste or utilized water. To practical ends, all water volumes used that are not reconditioned to give them immediate exploitation properties back constitute a reduction of the accessible water reserve that is usually compensated for with a substitution of new flows, which tend to put pressure on ground or underground availability of supply reserves.

Institutionally speaking, the social demand for water availability is usually met, under conditions of political comfort and technological stability, by gradually increasing the extraction of hydric resources with a small capacity to promote a transition in the way supply is run. This usually generates concentration of budget resources and efforts on extraction, transfer, and distribution activities. Thus, reducing the management capacity for maintenance activities in the carrying lines, which in turn causes a deterioration process and an eventual reduction of the transmitted flows due to fractures in the distribution lines.

This puts into perspective the fact that in the face of scenarios of relative institutional stability regarding incentives and rewards related to moderation in the consumption, with significant fiscal restrictions versus the operation costs, which would subsequently have to incorporate those related to the reestablishment of supply sources, and under conditions of persistent technological stability to meet the public problems of water supply, an inevitable result is the water shortage as a product of the imbalance in managerial and institutional architecture.

1.2 Water Management in Mexico and the Predominant Contemporary Model

Addressing water management in Mexico is not an easy task. In a decentralized and federal government regime there are several possibilities for approaching the issue related to various characteristics, such as the different tiers of government responsibilities facing the problem, the pressure put on supply sources, the response capacity for the cleaning of used waters, the existence of more or less reserves, the magnitude of ecological costs induced by increasing extraction, characteristics of social inequality, and the existence of heterogeneous forms of exploitation. However, in spite of all existing differences in the country related to the above-mentioned elements, there are some shared aspects that determine the exploitation trajectories in the country as a whole. Among others, the preponderance of the federal government over the other tiers of the institutional structure, as well as the existence of a strategy that has privileged water offer and led to exceeding exploitation of the reserves, in the face of a lack of incentives moderating the demand and inducing cautious use of the resources, can be highlighted.

The contemporary model of water management is the sum of a series of processes, some of which have maintained a relative stability over time and have given shape to an exploitation pattern that responds to the institutional design (Tortolero 2000). Without incentives for agricultural producers, the main users of the largest part of eater volumes, tend to the promotion of technical change that makes water use more efficient and, at the same time, arousing the idea of a more responsible water culture among domestic users via a price reassignment (Jiménez et al. 2010). A phenomenon that presents relative inconsistency in the water demand correction strategy by omitting incentives aimed at the agricultural sector. The result of all this cannot be other than an increasing pressure on the surface of underground water reserves.

There is an important institutional tradition regarding water management in Mexico, which goes back to the colonial times. Ever since then, it is possible to identify the moments over time in which the scarcity of water has been invoked as a managerial resource to introduce modifications that have tended to accumulate and prevail, sometimes in an unnoticed way, up to these days: first, regarding the supply volumes of surface bodies,. then in relation to the supply coming from underground sources.

Some cases that adequately illustrate the remarks about scarcity caused by institutional and managerial conditions are found in the historic water archive. For instance, the case of the claims received by the Mexico City council regarding the reduction of water supply flows coming from the Santa Fe aqueduct in 1869. The shortage had different explanations; first of all, illegal canalizations (Ávila 1997) being made to detour water toward private potable water fountains because of the lack of public fountains (Suárez/Birrichaga 1997), as well as the corresponding derivations of the aqueducts to increase the potency of the power feeding the nascent factories in the Mexico Valley.

The above is completed with the authority's response to the identified problems and the technical decision that avoided flow reduction in the face of the detour of water coming from the aqueduct meant the design of underground piping to avoid flow loss through canalizations. This resulted in the gradual reduction of supply in public fountains such as the Salto del Agua or the Candelaria de los Patos in old Mexico City.

The notion of water shortage is not a sign that is exclusive of contemporary Mexican society. The competition for water use for the attention of different purposes has marked the evolution of the hydric sector in Mexico over time. Some examples of this are the many documented cases of dispute between different types of users trying to achieve larger exploitation volumes (Suárez/Birrichaga 1997; Iracheta 2001), among which the conflicts regarding the use of the Magdalena River as it passed through Coyoacan in 1789 (Ávila 1997), the conflicts between the use for irrigation related to population supply (Von Wobeser 1993) and the one related to water as a source of hydraulic energy for motor force in factories from the eighteenth century on (Suárez/Birrichaga 1997) and the deviations of water in towns and villages (Ruíz 1986).

All of the former puts into perspective that the model for water management has, over the long term, evolved through a process of increasing offer, based upon the highly spread belief of an abundant natural richness in the country (Salmerón 2003); given the exploitation patterns among the different types of uses, reaching the physical limits of the infrastructure capacity, tensions generate and they tend to be labeled as caused by scarcity.

The period between the end of the nineteenth century and the first half of the twentieth century constitutes a foundational moment in the structure of the water management policy and the accumulation of imbalances that would later be the basis of most of the contemporary diagnoses. On the one hand, the gradual process of water federalization that corresponds with the introduction to agriculture of great scale irrigation through the promulgation of the Law of Water Exploitation at the Federal Level in 1910 (Aboites/Tena 2004). On the other hand, the increasing federal exposition in the management of potable water services in local contexts that allowed the consolidation of a culture of apparent gratuity in the supply service, not because that was its purpose, but given the lack of updates in prices and tariffs

Table 1.1 Water supply and economic growth

Variable	Coefficient	Standard Dev.	t-statistic	Prob.
C	1.408683	0.874285	1.611241	0.1176
Ln supply	1.007327	0.097454	10.33639	0.0000
R-square	0.780767	Dep. Variable Mean		10.41433
R-square adjusted	0.773460	Dep. Variable S.D.		0.864068
Regresion S.G.	0.411264	Inf Akaike criterion		1.121300
Quadratic sum of errors	5.074150	Schwarz criterion		1.212909
Verisimilitude Log	−15.94080	F-statistic		106.8409
Durbin-Watson statistic	1.405449	Prob (F-statistic)		0.000000

Isoelastic model LnGDP = f (Ln Supply). *Source* Own elaboration with data from: INEGI, National Account System. Conagua, Water Statistics (2011)

resulting from the lack of controls by the federal authorities.[1] Decentralization of operations, modernization of user records, and a new institutional architecture in the face of fiscal deficit, are attempts to revert this, from the decade of the 1980s.[2]

The evolution of the institutional architecture on the subject of water along the twentieth century consolidated a model of stable exploitation, in which water as a productive input shows constant returns to scale. Thus, part of the country's economic prosperity depends on a growing water volume (Table 1.1).

The idea that water can become an obstacle for the prosperity and compromise the welfare conditions of a society is an acceptable argument. However, if the reference to limits possibly imposed by its scarcity are associated only to the containing capacity of the infrastructure of the predominant sources for exploitation, different approaches should be explored that do not lead exclusively to an institutional response that increases the offer.

1.3 The Potential of Water Footprint and Virtual Water Approaches for Institutional Strengthening on the Subject of Water in Mexico

Scientific and technological advances on the subject of water have led to a better understanding of the natural processes determining availability through the water cycle. Additionally, they have led to a better idea of the institutional implications regulating the interchanges between the ecologic and socioeconomic systems, and the precursors of the apparent water scarcity.

---

[1]The promulgation of the Law of Cooperation for the Supply of Potable Water to Municipalities of 1956 is the institutional piece that would consolidate the federal intervention model in local management of potable water supply and the begging of the management deceleration process.
[2]The process starts with the reform to Article 115 of the Constitution, in 1983.

This has translated into a process of gradual strengthening of the technological capacities regarding supply, distribution, and quality assurance. Nevertheless, facing the increasing demand of water derived from demographic evolution, population relocation, economic performance, and use patterns, one of the elements that has been diagnosed as a potential contender for apparent water scarcity is the increase in the efficacy of water supply systems.

In the search for options to improve the country's water standpoint, it must be considered that the notion of scarcity that prevails currently is that which is linked to the limits of institutional supply strategy and not to the lack of efficiency itself. This is, ultimately, the visible result and not the main cause of institutional malfunctions over time.

The idea that public water affairs can be considered a transversal topic in a formal government agenda is closely linked to the implicit acknowledgment that water is a multidimensional factor of society's well-being. Everything needs water, one way or another, in an abundant list, but what can be summarized as the set of goods and services associated to the prosperity of a given society and its capability to reproduce itself: food, energy, production inputs, population's health, and manufactures of different complexities. Likewise, it is essential to the stability of ecological processes that accompany social and economic evolution.

The rise of alternative approaches to the water analysis within society such as those of water footprint and virtual water tend to emphasize the complex net of interconnections developed over the exploitation of hydric resources, through the estimation of the technical water coefficients required by the type of product, activity sector, or territorial region; and their eventual transformation into productivity and economic impact indicators. Therefore, the objective to develop a model of transversal management of water that does not contain information regarding supply and consumption volumes by type of source directly exploited only is strengthened through the incorporation of information about the economic impacts of contemporary exploitation patterns, their corresponding chains, and the potential water exploitation sources not commonly accounted for.

The concept of virtual water in the specialized literature usually refers to the volume of freshwater commonly required for the production of economic goods and services. Just as this literature points out (Garrido et al. 2010), the concept may have two different acceptations. The first , refers to the offer side of goods, as water required for the production of goods in their manufacturing or production site. The second , regarding the demand for goods, as the water content needed for the production and disposal of merchandises in the consumption site.

Just as the pioneer study by Arreguín (2007) asserts for the Mexican economy, as well as the converging study of Aldaya (2008) for the international case, commerce of goods between regions within a country or at world level allows contending the negative effects of limited water availability which may affect, among other things, food production. Goods and services interchange is, in a way, a process of exchange and moving of water. But not an exchange or move of the water directly contained in them, but of that required for their production and disposal (Mekonnen/Hoekstra 2011).

Along with the concept of virtual water, the notion of water footprint or hydrologic footprint has risen with increasing importance. This is an indicator of the social appropriation of freshwater resources employed in the production of the flow of goods and services, apart from those contaminated by time unit. Water footprint has three components called blue water, green water, and gray water. Blue water is identified as the superficial or underground water reserves that are usually the object of traditional water policy through the construction of infrastructure that makes their exploitation immediate. On the other hand, green water is the component of the hydrologic cycle, also known as non-saturated zone water which allows the existence of vegetation based on soil humidity (Aldaya 2008; Garrido 2010; Mekonnen/Hoekstra 2011). Gray water represents the volume of water that is polluted during the production and consumption processes.

A region or a country's water footprint is estimated from the account of exploitation of local water resources and the incorporation of virtual water resulted from importing goods. In an evolution process of this approach, known as extended water footprint, the socioeconomic impacts of the water exploitation pattern are usually added by linking utilized volumes and the value of production or the induced effect in occupation growth, as well as the potential ecologic impacts.

From a perspective such as that of the water footprint and virtual water approaches, reflections on water scarcity, competition for the exploitation of hydric resources, and the impacts on public welfare of how those resources are utilized, take on a different dimension.

In Mexico, the idea of scarcity is usually associated to the limits of blue water exploitation, both in terms of the limits imposed by storage capacity of the infrastructure developed in the country for the direct exploitation of such resources, as well as in terms of the financial restrictions imposed by a structure of increasing costs derived from the exploitation pattern fostered over time, which has meant increasing exploitation of the supply sources, a frequent practice of transfers between basins, the increase of intangible costs ecology wise due to the prolonged deterioration of ecosystems, and an incentives structure limited to correct water consumption patterns (Jiménez et al. 2010).

There is a gradual institutional tendency in the country toward recognizing that water issues require a transformation in the way they are approached. Some indications of that can be found dispersed in the 2030 Water Agenda (Conagua 2012), the Special Program on Science and Technology on Water (FCCyT 2012) or the Special Concurring Program for the Sustainable Rural Development (CISDR 2007), just to mention a few of the contemporary government documents processing objectives aimed at reaching water sustainability. Nevertheless, the transversality of water issues demands harmonization processes for economic policy for the generation of incentives that correct the inefficient uses of water, the industrial and rural development policy that promotes exploitation processes with the highest possible economic impact, and the social policy tending to attenuate inequality among the population.

There is growing concern at the international level for the effects on development and welfare derived from the links between water and food production in a context of major climatic swaying (Waughray 2011). Mexico, just a like a great number of countries, devotes an important portion of its water extractions for the production of the agricultural sector. Arreguín (2007) has shown, based on his analysis of virtual water, that in the primary sector an involuntary result on water is that the Mexican economy is a net importer of virtual water based on the commerce of food. The best knowledge of water exploitation according to its colors and potential economic effects is a central factor for the construction of an institutional regime that promotes the best practice that increases resilience against meteorological changes in a territory in which drought tends to affect the population recurrently (Semarnat 2011).

Of course, and as Andre Santos states in Chap. 9 of this book, the fact that the country reduces its requirements of water coming from local sources for the production of food does not necessarily mean that an adequate food model is guaranteed.

The presentation above puts into perspective that there are a set of advantages in the utilization of the water footprint and virtual water approaches, but also limitations. The advantages are, the fact that they allow to make sense of the idea of required transversality in the implementation of the country's water policy must be highlighted. Second, they allow to identify a priority regime for the design of incentives based on the different colors of water for the development of the corresponding infrastructure. Third, they make it easier to articulate commercial strategies in the external sector that reduce pressure on local water resources.

Among the most significant restrictions identified in a transition process that strengthens water management policy upon the basis of the water footprint and virtual water approaches, are a lack of appropriate information systems for the estimation of technical coefficients according to water color and the adequate volumes consumed by product or region. Likewise, the presence of power groups that may be an obstacle for initiatives of change is important, due to the fact that these may mean they have to face significant costs derived from a turn in the conduction of water policy.

Of course improving the country's position upon the basis of the water footprint and virtual water analysis does not imply that the potential increase in welfare means a reduction of social inequality or that the food strategy is the fittest. Such demands cannot be met by water policy.

References

Aboites, Luis and Valeria Estrada. *Del agua municipal al agua nacional. Materiales para una historia de los municipios de México (1901–1945)*. México: CONAGUA/AHA/CIESAS/COLMEX, 2004.
Aldaya, Maite, Ramón Llamas, Alberto Garrido, and Consuelo Varela. "Importancia del conocimiento de la Huella Hidrológica para la política española del agua", *Encuentros Multidisciplinares 10* (2008): 1–9.

Arreguín, Felipe, Mario López, Humberto Marengo and Carlos Tejeda. "Agua virtual en México" in *Ingeniería Hidráulica en México,* vol. XXII, no. 4, (2007): 121–132.

Ávila, Salvador. *Guía de fuentes documentales para el estudio del agua en el Valle de México (1824–1928).* Mexico: Archivo Histórico del Ayuntamiento de la Ciudad de México/IMTA/CIESAS, 1997.

CIDRS. *Programa especial concurrente para el desarrollo rural sustentable 2007–2012,* México: Comisión Intersecretarial para el Desarrollo Rural Sustentable, 2007.

CONAGUA. *Estadísticas del agua en México,* México: SEMARNAT, 2011.

CONAGUA. *Agenda del agua 2030. Avances y logros 2012,* Mexico: SEMARNAT, 2012.

FCCyT. *Programa Especial de Ciencia y Tecnología en Materia de Agua,* México: CONAGUA/IMTA/CONACYT, 2012.

Garrido, Alberto, Ramón Llamas, Consuelo Varela, Paula Novo, Roberto Rodriguez and Maite Aldaya. *Water footprint and virtual water trade in Spain. Policy Implications.* USA: Fundación Botín, Springer, 2010.

INEGI. *Sistema de Cuentas Nacionales,* México: INEGI, 2011–2012.

Iracheta Cenecorta, María del Pilar. "El aprovisionamiento de agua en la Toluca colonial", *Estudios de Historia Novohispana* 25 (2001): 81–116.

Jiménez, Blanca, Maria Luisa Torregrosa y Armentia and Luis Aboites. *El agua en México: cauces y encauces,* Mexico: Academia Mexicana de Ciencias, 2010.

Mekonnen, Mesfin and Arjen Hoekstra. *National water footprint accounts: The green, blue and grey water footprint of production and consumption* 1, Main Report, UNESCO/IHE/Institute for Water Education, 2011.

Ruíz Gomar, Rogelio. "La fuente de la antigua plaza del santuario de Guadalupe", *Anales del Instituto de Investigaciones Estéticas,* vol. 15, no. 57, (1986): 75–98.

Sainz, Jaime and Mariana Becerra. "Los conflictos por agua en México", *Gaceta Ecológica 67* (2003): 61–68.

Salmerón, Pedro. "El mito de la riqueza de México. Variaciones sobre un tema de Cosío Villegas", *Estudios de historia moderna y contemporánea de México,* vol. 26, no. 26, (2003): 127–152.

SEMARNAT. *Análisis espacial de las regiones más vulnerables ante las sequías en México,* México: SEMARNAT, 2011.

Suárez Cortez, Blanca E. and Diana Birrichaga. *Dos estudios del agua en México (siglos XIX y XX),* México: IMTA/CIESAS, 1997.

Tortolero, Alejandro. *El agua y su historia. México y sus desafíos hacia el siglo XXI. Umbrales de México,* México: Siglo XXI Editores, 2000.

Von Wobeser, Gisela. "El agua como factor de conflicto en el agro novohispano, 1650–1821", *Estudios de Historia Novohispana* 13, (1993): 135-146.

Waughray, Dominic. *Water security: The water-food-energy-climate nexus. The World Economic Forum Water Initiative,* USA: World Economic Forum, Island Press, 2011.

Chapter 2
Socioeconomic Framework

Hilda R. Dávila-Ibáñez, Rosario H. Pérez-Espejo
and Thalia Hernández-Amezcua

Abstract The use of the water resource is closely related to the geographical situation of the territory, but above all to the type and degree of socioeconomic development of countries, which is why in order to analyze water footprint in Mexico it is necessary to put the country's characteristics into context. This chapter presents a general view of the country and puts it into the World and Latin-American contexts based on the main socioeconomic indicators, highlighting the unequal income distribution, the rural bias of poverty and the particular importance of the agricultural sector. This sector's heterogeneity, as well as its relevance as a water user and an important cause of its deterioration are mentioned. We also mention that notwithstanding the importance of irrigation, to a good extent with underground water, the most important crops in Mexican people's diet is produced in rain-fed lands, and a high percentage of both irrigated and rain-fed fields are devoted to livestock raising. Mexico's growing dependence on food imports and the paradox that this brings environmental benefits is also commented upon.

Keywords Socioeconomic indicators · Water · Agriculture

2.1 Geographic and Demographic Characteristics

Mexico is country of contrasts and inequalities, both geographic and socioeconomic. In its territory spanning 1,964,375 km^2, located between meridians 118° 22′ and 86° 42′ east and latitudes 14° 32′ and 32° 43′ north, there is a great variety of climates. Two thirds can be considered arid or semiarid with precipitations below 500 mm annually and a third part located in the southeast with precipitations over 200 mn a year. Additionally, according the the National Water Commission (Spanish acronym: Conagua), 53 % of the population live in levels of over 1500 m of altitude (Conagua 2012). The 2010 Population and Housing Census (INEGI 2010) records 112,336,538 inhabitants. The National Population Council (Spanish acronym: Conapo) estimates this figure will reach 118,395,054 inhabitants by mid 2013.

© The Author(s) 2016
R.H. Pérez-Espejo et al. (eds.), *Water, Food and Welfare*,
SpringerBriefs in Environment, Security, Development and Peace 23,
DOI 10.1007/978-3-319-28824-6_2

Since 1950 the country has undergone an accelerated urbanization process. It has gone from a rural majority (57.3 %) to become an urban country (76.8 %) this urban population is concentrated (29 %) in five great metropolitan areas: the Federal District, Mexico State and Hidalgo; Guadalajara; Monterrey, Puebla, Tlaxcala and Toluca. According to Conapo estimations this concentration will continue over the following years, which will increase the demand for public services in these regions.

Geographic and demographic contrasts are also reflected in the socioeconomic characteristics of the population. According to the marginalization study conducted by Conapo with census information,[1] in spite of the advances mad in marginalization indicators from different places the situation is still worrying as it can be seen in Fig. 2.1 There are 441 localities with a very high degree of marginalization, most of which are located in Guerrero, Chiapas, Oaxaca, Veracruz and Puebla.

Looking at the population welfare levels through the indicators of the United Nations on Human Development,[2] elaborated by the United Nations Development Program (UNDP) in Mexico, we see a similar situation. The development index for 2011 places Mexico in place 57 in the international list with a value of 0.770, that is, within the 25 % with high development. However, inequality among the different entities is acute. The highest welfare indexes are found in the Federal District, Nuevo Leon and Baja California, and they can compare to those in the Czech Republic or Poland. In contrast, Chiapas, Oaxaca and Guerrero have the lowest welfare indexes for the country (UNDP 2012b) (Fig. 2.2).

The other side of the coin is poverty among great sectors of the national population. Based on information from 2012, 45.5 % of the population was in poverty, which represents 53.3 million people, out of whom 11.5 million are in extreme poverty, equivalent to 9.8 % of the national population. Most of this population is found in Chiapas (1,629,200), Veracruz (1,122,000), Guerrero (1,111,500), Puebla (1,059,000), Mexico State (945,000) and Oaxaca (916,000).

---

[1]For the confection of its marginalization index for localities, Conapo considers the following variables: percentage of illiterate population 15 years old or more, percentage of the population without elementary school studies 15 years old or more, percentage of population living in housing no sanitary services and drainage, percentage of people living in housing without electric power, percentage of people living in housing without water piping, percentage of housing with overcrowding, percentage of people living in housing with dirt floor, percentage of population in localities with less than 5,000 inhabitants, percentage of people with income below two minimum wages.

[2]"The Human Development Index synthetizes the average progress on three basic aspects of human development, measured in a zero to one range, in which values closer to one represent higher human development. In reports previous to the twentieth edition of the HDI, the long and healthy life used to be measured by the life expectancy at birth index; the access to knowledge index was obtained by using the literacy rate and the combined enrollment rate together; while the decent standard of living was calculated through the gross domestic product per capita in Purchasing Power Parity (PPP) stated in US dollars. Thus, the HDI was obtained as the simple average, or arithmetic mean, of those three indicators".

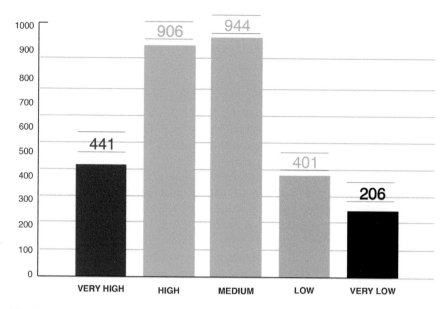

Fig. 2.1 Municipalities by degree of marginalization. *Source* Own elaboration with data from INEGI (2010)

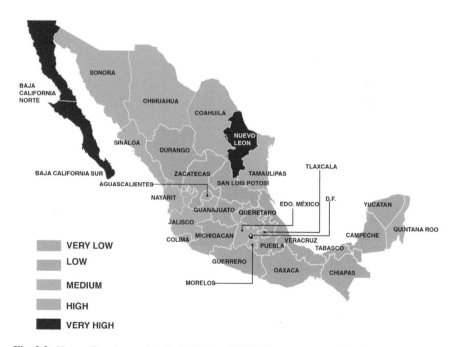

Fig. 2.2 Human Development Index in Mexico (2011). *Source* cartographic elaboration with the support of the University program for Metropolitan Studies, UAM, with data from UNDP (2011)

The 2008–2009 crisis was reflected in an acute increase of all type of poverty levels for 2010, which has not been able to be reverted in reverted in recent years. According to the new multidimensional methodology for the measure of poverty in Mexico, elaborated by the National Council for the Evaluation of Social Development Policy (Spanish acronym: Coneval), a concept that includes the variables income and social deprivations and an income below the welfare line. The population in poverty went down from 46.1 to 45.5 %, although it actually increased in number of people from 52.8 to 53.3 million; the population in extreme poverty— with an income less than the minimum welfare line and 3.7 in social deprivations— was decreased from 13 to 11.5 million people between 2010 and 2012. Although on the other hand there was a decrease in the real income of households, especially in urban areas, which resulted in an increase in the population with an income below the minimum welfare. The alarming thing is that only 19.8 % of the total population may be considered as not poor and not vulnerable (Fig. 2.3).

Poverty has a rural bias. In zones with a population of less than 2,500 inhabitants, the poverty rate is much higher than in urban zones, although it must be noted that there is a tendency in the opposite direction, the transfer programs such as *Oportunidades* (opportunities), which benefit rural areas as a priority, and the economic crisis of 2008–2009 that affected urban zones in a greater way, have increased poverty among cities' population (Fig. 2.4).

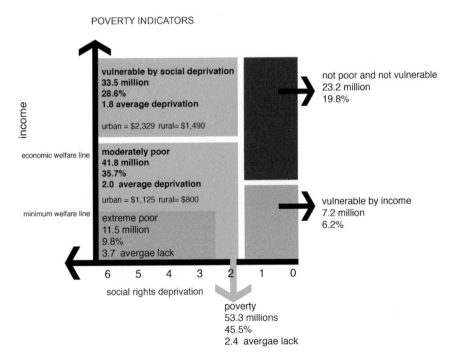

Fig. 2.3 Poverty indicators 2012. *Source* Own elaboration with data from Coneval

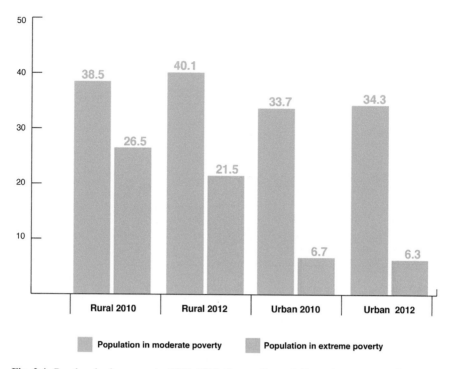

Fig. 2.4 Rural and urban poverty, 2010–2012. *Source* Coneval, Executive summary (2013)

Poverty and inequity in Mexico do not only manifest themselves among the different federal entities, they are also patent within them, for instance, income inequality among households was 0.453 at the national level. Being income distribution in Mexico one of the most inequitable in Latin America, a region marked as the most unequal in the world. In 2012, 30 % of households in the last three deciles of income distribution concentrated 56.3 % of the national current income; while the remaining 70 % only had 43.7 % of the income. 10 % of the most favored population concentrated 30.1 % of the income, while 30 % of the poorest population only had 11.9 % (ENIGH 2012) (Fig. 2.5).

2.2 Economic Distribution of the Population

In Mexico, economically active population (EAP) is 50.2 million people and it accounts for a 43.9 % of the total population; 13.3 % of the EAP is the primary sector (around 6 million people); 22.2 % is occupied in the secondary sector and 59 % in the third sector, mostly as informal employment. The unemployment rate was estimated at 4.8 % (INEGI 2012a, b).

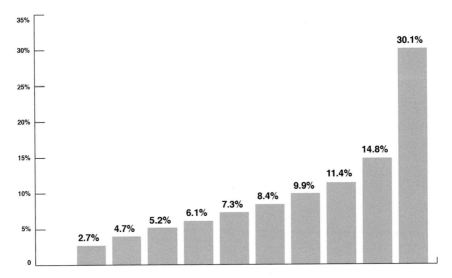

Fig. 2.5 Income distribution by deciles. *Source* own elaboration with data from the ENIGH 2012

The distribution of the economically active population, formerly mentioned, does not have a direct counterpart in product distribution. While the primary sector contains 13.7 % of the population, it only generates 4 % of the national gross domestic product, this is the result of the great differences in productivity that still persist between the primary sectors and the industrial and service sectors (Fig. 2.6).

Fig. 2.6 GDP composition and employed population by sector, 2009. *Source* INEGI (2009)

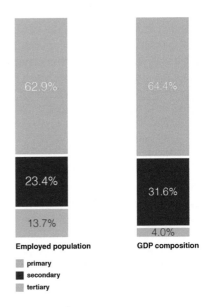

A similar phenomenon takes place between population distribution, economic development and the amount of renewable water; thus, in the north, center and northeast part where 76.9 % of the population is concentrated and 78.9 % of the generated domestic product, only 31.74 % of renewable water can be found, versus the remaining 68.2 % which is found in the south and southeast parts of the country.

2.3 Conclusions

Mexico is a country of contrasts and inequalities in virtually every category: geographic, climatic, environmental and hydric; these inequalities go together with social and economic inequality between federal entities and within each of them.

There is a contradiction between development and water distribution; the states with the lowest level of income are those that have the largest quantity of hydric resources, coming fundamentally from precipitation. Nevertheless, in spite of the fact that a large portion of the population is employed in agricultural activities, they are not the main farm producers at the national level. Chiapas, Oaxaca and Guerrero present the highest marginalization and poverty indexes in the country, while they also have the biggest quantity of hydric resources.

The regions that now present the highest poverty indexes and the lowest in development, have enormous potential for development for alimentary security, given the human and hydric resources they possess.

References

CONAGUA. *Atlas del agua*, México: SEMARNAT, 2012.
CONEVAL. http://www.coneval.gob.mx, Accessed January 19, 2014.
CONAPO. *Índice de marginación por entidad federativa y municipio 2010,* México: SEGOB, 2010.
Cortés, Fernando. *Desigualdad económica y poder en México*, México: CEPAL, 2011.
INEGI. *Encuesta nacional de ingresos y gastos de los hogares,* México: INEGI, 2012.
INEGI. *Censo nacional de población y vivienda 2010*, México: INEGI, 2012.
UNDP. *Índice de desarrollo humano en México 2012*, México: ONU, 2012.

Chapter 3
Water Policy and Institutions

**Rosario H. Pérez-Espejo, Thalia Hernández-Amezcua
and Hilda R. Dávila-Ibáñez**

Abstract This chapter exhibits Mexican legislation on the subject of use and exploitation of Mexican water resources at the different government tiers and bodies, and lays out a scheme of water policy based on its main instrument, the *Plan Nacional Hídrico* (National Water Plan), in which the water management by basin and the social participation in decision making, the latter a more expositive principle than real. In spite of the fact that Mexico has a reasonably adequate legal framework on water subjects, with a set of institutions, among which Conagua stands out, and a water policy whose instruments have diversified, water management presents a series of problems such as a lack of long-term view, a bias toward farming water use, and the development of hydro-agricultural infrastructure and budget allocation that neglects sanitation, sewage, and water quality needs.

Keywords Regulatory framework · Institutions · Water management

3.1 Water Legal Framework and Institutions

According to Article 27 of the Political Constitution of the United Mexican States, the property of waters found within the national territory corresponds originally to the nation, which has the right to regulate its exploitation and transfer its control to particulars. Water use and exploitation is done through concessions granted by the Executive Branch of the Federal Governments.

According to the National Water Law (NWL) of 1992, modified in 2004, water is a public good, a strategic resource whose management is a national security issue (Article 14). Since 1994, the Federal Executive Branch exercises water authority and administration through the National Water Commission (Spanish Acronym: Conagua), a decentralized agency of the Secretariat of Environment and Natural Resources (Spanish acronym: Semarnat).

The Federation norms, plans, manages, and operates the resource water and the municipalities are responsible for administering potable water services, sewage, and treating residual waters (Constitution, art. 115-I). The Federation collects the rights for water exploitation and wastewaters disposal, as well as fines and late fees

© The Author(s) 2016
R.H. Pérez-Espejo et al. (eds.), *Water, Food and Welfare*,
SpringerBriefs in Environment, Security, Development and Peace 23,
DOI 10.1007/978-3-319-28824-6_3

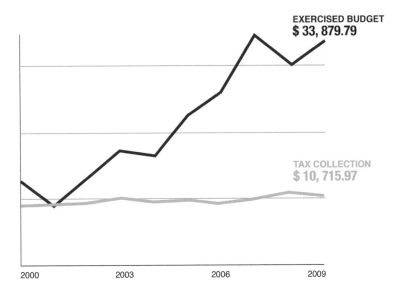

Fig. 3.1 Tax collection and budget spent by the national water commission. *Source* Conagua 2011

generated by the former concepts. This collection (Fig. 3.1) comes from only 5 % of water users due to a generalized no payment culture and many operating organizations—especially in those states where the resource is scarce and has a high cost—do not pay the corresponding fees. Low income, derived from insufficient tariffs and scarce payment, generate a perverse circle in which users do not pay for the service because they consider it unfit and operating agencies do not offer a better service because of the lack of payment. Exploitation oriented to farming use (78 % of the extracted water) are not the subject of payment (Guerrero 2004: 31–46).

The 2002 reforms of the NWL included the world tendencies in water management: the basin as the management unit, payment for water consumption, the principle that "he who pollutes has to pay," the acknowledgment that there is a need for integral and integrated water management, social participation in management, and decision making at the level problems present themselves. Progress has been made regarding federalism, decentralization, and administrative disaggregation; some responsibilities were delegated to states and municipalities and there is a possibility to establish coordination agreements for solution of specific issues (Carabias/Landa 2005). But municipalities were not given the necessary resources to fulfill their new responsibilities.

Conagua does not only exercise water authority and management, it also takes actions on the vigilance over the resource that are not included in the responsibilities of the Federal Attorneys Office for the Protection of the Environment (Spanish acronym: Profepa), which is in charge of watching over maritime resources, federal sea–land zones and maritime waters, leaving underground and surface waters supervision to the Conagua.

State offices, whose jurisdiction are state waters less important than federal waters, were added to the federal structure for water management represented by Conagua. Thus, every state of the federation, as well as the Federal District, has a state water commission regulated by a state water law.

The provision of potable water, sewage, and wastewater treatment services is carried out through operational agencies,[1] 2,517 units in 2008 (INEGI 2011) that can be public or private and are mostly located in urban areas.

To the end of water management, Mexico is divided into 13 hydrologic-administrative regions (HAR); each of these has a basin agency (BA) reporting to Conagua's general director, and has the same functions as Conagua, only at a regional level. The BA are formed by a general director, an advisory council, a representative from the state's Federal Branch and another one from the municipalities within the BA's jurisdiction, as well as a users' representative.

In a parallel way and not subordinated to the structure integrated by Conagua, the BA, and the state commissions, there are other agencies called basin councils (BC), with a mixed integration and that have coordination, agreement, support, consulting, and advisory functions they display between government structures (Conagua, water federal, state or municipal agencies and entities) and the HAR's users' representatives or organizations. The BC work as a General Assembly of Users, a Directive Committee, an Operation and Vigilance Commission, and an Operative Management office. They are supported by basin commissions (in sub-basins or groups of subbasins), basin committees (micro basins or groups of micro basins), and the technical committees of underground waters (TCUW).

In the organizational structure of water management (Fig. 3.2), we can observe that Conagua is attached to Semarnat, of which it accounts for nearly 70 % of its budget. On the other hand, directly subordinated to the director of Conagua we have the National Meteorological Service (Spanish acronym SMN), a specialized autonomous technical unit, and the Mexican Institute of Water Technology (Spanish acronym: IMTA), a decentralized public organization that is the first technological advisor to Conagua.

3.2 Water Policy

The Head of the Semarnat submits a proposal for the national water policy and the *National Hydric Plan* formulated, updated, and watched over by Conagua, to the Executive Branch of the Federal Government (NWL, Articles 8 and 9). According to the NWL (Articles 14 BIS, 5 and 6), water policy and program are based upon a series of basic principles among which the most important ones are: (a) water is a

---

[1]Operating agencies are usually part of municipal governments and are represented by potable water and sanitation commissions and offices, or decentralized water systems. They also operate as local water users committees or associations, and, less frequently, as private enterprises with concessions (INEGI 2011).

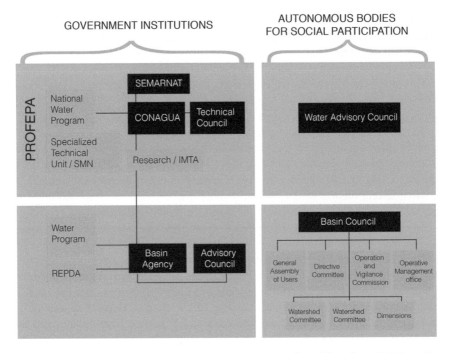

Fig. 3.2 Organizational structure of water management. *Source* Own elaboration with information from Conagua

public, vital, vulnerable, and finite resource, with social, environmental, and economic value, the preservation of which is a matter of national security; (b) attention to marginalized population; (c) payment for water exploitation or use; and (d) domestic and public uses are a priority (Annex 3.1).

According to that same Article in the NWL, water policy and program are also supported by eight instruments, among which we find: (a) water planning at the different geographical levels; (b) concessions, assignments, and permissions rules; (c) the collection of fees; and (d) the social supports for access to water and sanitation (Annex 3.2).

Water planning, the main instrument of water policy and program, contains nine elements, of which the most important one is the *National Water Program*. Water programs at different space levels: basin, state, aquifer; special and emerging programs, subregional programs by BCs' Advisory Councils (Article 12 bis 6 of the NWL), and so on, are also part of this planning (Annex 3.3).

BCs, in coordination with BAs, propose to Conagua the preference order for water uses, in which, in accordance to water policy, domestic and urban public use are given prioritized. In Article 15 transitory of the NWL, there appears a preference order that aggregates livestock and agricultural use as priorities (Annex 3.4).

The *National Water Program 2007–2012* (NWP 07-12) had water management by hydrological basin and social participation in decision making as basic principles. The NWP 07-12 included eight guiding objectives (Annex 3.5), from which water productivity in agriculture occupies the first place, a place that does not correspond to the preference of uses established in the National Water Law itself.

There is a set of strategies for each of the eight guiding objectives of the NWP 07-12, with its accompanying indicators and goals; in spite of that, the *National Water Program* is still a very general instrument not specifying economic sectors, geographical spaces of specific instruments for the achievement of its objectives, and implementation of its strategies.

At the most disaggregate level of water policy, there is a set of Mexican Official Norms (Spanish acronym: NOM) which are mandatory on the subject of environment and natural resources.[2] On the subject of water, Semarnat has issued seven NOM. Three of them about water quality; directly, Conagua has issued 14 and the group of the Secretariat of Health, 6. Additionally, six Mexican Norms (Spanish acronym: NMX) of voluntary observance, have been published in order to regulate varied aspects related to water.

The *2030 Water Agenda* elaborated by Conagua in March 2011, including 13 technical studies analyzing alternatives for sustainable water use by 2030, was added to the different elements of water policy. The set of proposed initiatives and actions are in tune with international guidelines laid out at different world water forums, which focus on balanced basins, clean rivers, universal coverage of potable water and sewage services, and the attention to climate change catastrophic impacts.

In spite of the fact that Mexico has a reasonably adequate legal framework on water, a group of institutions among which by regulating, running, and watching over water management, Conagua stands out, water policy with diversified instruments including citizen participation, transparency, and accountability, the following problems in water management can be observed:

1. A lack of long-term vision: sector's policy is formulated and modified every six years with the change of administration;
2. A bias toward agricultural use of water and the development of hydro-agricultural infrastructure persists, and is still present in the budget allocation that neglects sanitation, sewage, and water quality needs;
3. A water planning system and the possibility for social participation that is more formal than real;
4. A lack of political will and resources to watch over the sector's regulations; and
5. An inefficient water management that is not solved with higher budgets or applying tariffs, but requires a profound evaluation of priorities and budget exercise.

---

[2]They are also mandatory on sanitary, labor, and security issues.

Annex

Annex 3.1: Basic principles of water policy and programming. *Source* The authors.

1. Water is federal public, vital, vulnerable and finite good, with social economic and environmental value and its preservation and sustainability are a priority and national security issues;
2. National water policy based on the integrated and decentralized management of hydric resources by hydrologic basin that privileges local actors' decisions;
3. Attention to the water needs for welfare, development, and sustainability. Marginalized population is a priority;
4. The State regulates water uses in basins, aquifers, and transfers. Water concessions and assignments shall take its availability into account;
5. Unsustainable use of water will be avoided and its interrelations to other natural resources vital for water will be taken into account;
6. Environmental services provided by water must be acknowledged, quantified, and paid for, and its reuse must be promoted;
7. Measures for appropriate water quality for human consumption shall be taken in order to have an impact on public health;
8. Water users must pay for its exploitation, use, or utilization;
9. Those who pollute water shall restore its quality, the principle "whosoever pollutes, must pay" shall be applied and there shall be economic incentives for its efficient and clean use;
10. Education on the subject of water shall be essential;
11. Domestic and public urban use shall have preference.

Annex 3.2: Basic instruments of water policy and programing. *Source* The authors.

1. Water planning at different geographic levels;
2. Concessions, assignments, and permissions (for water exploitation, use or utilization; use of national goods and discharge permissions) regulation;
3. National waters management;
4. Collection of fees (for exploitation, use, or discharge);
5. Social participation;
6. The resolution of conflicts on the subject of water (prevention, conciliation, mitigation);
7. Social supports (for access to water and sanitation);
8. The National Water Information System.

Annex 3.3: Water planning elements. *Source* The authors.

1. The *National Water Plan* (six-year period);
2. Water programs by hydrologic basin or group of basins;
3. Specific subprograms by region, hydrologic basin, aquifer, state, or sector;
4. Special or emergency programs;
5. Integration and updating of the catalog of water exploitation or utilization programs, and those for its preservation and quality control;
6. Classification of water bodies according to the use they are devoted to, and the elaboration of water balances in quantity and quality, as well as by basin, HAR, and aquifer, according to their own capacities;
7. Strategies and policies for the regulation of water exploitation, use, or utilization and its preservation,
8. Mechanisms for consultation, agreement, participation, and the taking on of specific commitments for the realization of programs and their financing;
9. Multiannual investment programs and annual operative programs for investment and action by the National Water Commission (Spanish acronym: Conagua)

Annex 3.4: Preference of water uses in Mexico. *Source* The authors.

1. Domestic;
2. Public urban;
3. Cattle and livestock raising;
4. Agricultural;
5. Ecologic preservation or environmental use;
6. Electric power generation for public service;
7. Industrial;
8. Aquaculture;
9. Electric power generation for private service;
10. Land washing and sliming;
11. Tourism, recreation, and therapeutic purposes;
12. Multiple uses;
13. Others

Annex 3.5: Guiding objectives of the *National Water Plan. Source* The authors.

- Improve water productivity in the agricultural sector.
- Increase access and quality in potable water, sewage, and sanitation services.

- Promote integrated and sustainable water management in hydrologic basins and aquifers.
- Improve technical, administrative, and financial development of the water sector.
- Consolidate users and organized society's participation in water management and promote the culture of good water use.
- Prevent risks sprung from meteorological and hydro-meteorological phenomena and meet its effects.
- Evaluate the effects of climate change in the water cycle.
- Create a contributing and of National Water Law abiding culture.

References

* indicates internet link (URL) has not been working any longer on 8 February 2016.

Carabias, Julia and Rosalva Landa. *Agua, medio ambiente y sociedad. Hacia una gestión integral de los recursos hídricos en México*, México: UNAM/COLMEX/Fundación Gonzalo Río Arronte, 2005.
CONAGUA. *Programa Nacional Hídrico 2007–2012*, México: SEMARNAT, 2008.
CONAGUA. *Estadísticas del agua en México*, México: SEMARNAT, 2011.
*CONEVAL. *Pobreza 2010. Porcentaje de la población en pobreza según entidad federativa*, 2010, http://www.coneval.gob.mx/cmsconeval/rw/pages/medicion/pobreza_2010.es.do, Accessed on April 4, 2012.
Cortés, Fernando. *Desigualdad económica y poder en México*, México: CEPAL, 2011.
FAO. "*La FAO en México, más de 60 años de cooperación 1945–2009*" www.fao.org.mx/documentos/Libro_FAO.pdf, Accessed April 24, 2012.
Guerrero, Vicente. "Aportes de la gestión integral del agua". In *Hacia una gestión integral del agua en México: retos y alternativas*, edited by Cecilia Tortajada, Vicente Guerrero and Ricardo Sandoval, 31–46, México: Centro del Tercer Mundo del Manejo del Agua, México/Miguel Ángel Porrúa, 2004.
INEGI. "*Panorama censal de los organismos operadores de agua en México: Censos Económicos 2009*", http://www.inegi.org.mx/prod_serv/contenidos/espanol/bvinegi/productos/censos/e conomicos/2009/agua/Mono_Orgs_operadores_agua.pdf, Accessed April 24, 2012.
INEGI. *Perspectiva estadística*, México: INEGI, 2011.
*INEGI. *Encuesta nacional de ocupación y empleo. Indicadores estratégicos,* http://www.inegi.org. mx/inegi/contenidos/espanol/prensa/Boletines/Boletin/Comunicados/Indicadores%20estructu rales%20de%20ocupacion%20y%20empleo/2012/febrero/cuadro.xls, Accessed April 4, 2012.
Procuraduria Agraria. *Estadísticas agrarias*, México: Procuraduría Agraria, 2011.
Sistema de Información Agroalimentaria y Pesquera. *Agricultura. Producción agrícola,* http://www.siap.gob.mx/index.php?option=com_content&view=article&id=42&Itemid=12, Accessed April 25, 2012.
Sistema de Información Agroalimentaria y Pesquera. *Agricultura. Producción agrícola* http://www.campomexicano.gob.mx/portal_siap/Integracion/EstadisticaDerivada/ComercioExterior/Estudios/Perspectivas/maiz96-12.pdf, Accessed May 25, 2012.

Chapter 4
Water Use for Food Purposes

**Rosario H. Pérez-Espejo, Thalia Hernández-Amezcua
and Hilda R. Dávila-Ibáñez**

Abstract This chapter ponders how the agricultural sector uses water for food in
Mexico. It is based on a review of the geographic, social, and political conditions
under which the Mexican countryside produces, emphasizing its heterogeneity and
the presence of an external water footprint due to the importation of agricultural
products, cereals are the products with a higher content of virtual water imported by
Mexico, and although this, on the one hand, saves water, reduces soil loss and water
pollution because of a lower use of agrochemicals; on the other hand, food security
and sovereignty are undermined. Farming and livestock raising activities are the
main users of the water and soil resources, and are the most important causes for
their deterioration; altogether, these sectors use 78 % of extracted water: 76 % by
agriculture and 2 % for the cattle industry. Notwithstanding the importance of areas
on irrigation, the production of basic foods in Mexico, corn, beans, and wheat,
depends largely on the production from rain-fed zones. Only eight states were
found to contribute to the national farming and ranching producing a higher water
percentage than the amount they have in concession for agricultural use, and the
rest of the federative entities take hold of a higher percentage of the resource in
relation to what they provide in agricultural product.

Keywords Water · Agriculture · Agroindustry · Food production

4.1 Agriculture, Agroindustry, and Water Use

In Mexico, the agricultural and cattle sector is extremely heterogeneous and con-
trasting; it is an important stronghold of extreme poverty which expels population
toward the peripheral urban zones in the country and abroad. The average age
among rural population surpasses 52 years and a growing number of productive
units are now in the hands of women (Agricultural Attorney's Office, 2011).
Farming and livestock raising activities are the main users of the water and soil
resources, and are the most important causes for their deterioration. Of the almost
200 million hectares (Mha) in the country, 30 million have agricultural potential;
however, sown land has ranged between 20 y and 23 Mha in the last 10 years

R.H. Pérez-Espejo et al. (eds.), *Water, Food and Welfare*,
SpringerBriefs in Environment, Security, Development and Peace 23,
DOI 10.1007/978-3-319-28824-6_4

(Conagua 2012). In 2010, sown land was almost 22 Mha and harvested land was 20 (SIAP 2012a). Cattle raising occupies the majority of the national territory, around 112 Mha. Farming and ranching sectors together use 78 % of the extracted water: 76 % for agriculture and 2 % for livestock raising. If the amount of water consumed by crops destined to cattle feeding were added to the sector's accounts, its water consumption would be much higher. Notwithstanding the enormous amount of natural resources these sectors use, they only contribute 3.2 % of the gross domestic product (GDP). However, it employs 18.3 % of the economically active population (EAP) (around six million people) on which about a quarter of the total population depends.

Because of the magnitude of its irrigation surface, Mexico is in the sixth place with 6.46 Mha, of which 54 % corresponds to 85 irrigation districts and the rest to more than 39 thousand small irrigation units for rural development (Spanish acronym: Urderal) about which there is little information available. 33.8 % of water assigned as concessions for agriculture, aquaculture, cattle raising, and other multiple uses has an underground origin, and water use efficiency in agriculture is 46 % (Conagua 2012).

In 2010, rain-fed sown land was 16.2 Mha and the harvested land accounted for 14.6 Mha; irrigation land had the following figures: 5.6 Mha (sown) and 5.5 Mha (harvested). Risks associated to climate changes and extreme hydro-climatic events caused that 10 % of the rain-fed land and 3 % of the irrigated land was not harvested. Notwithstanding the importance of irrigation areas, basic foods production in Mexico—corn, beans, and wheat—depends to a great extent on rain-fed zones. In 2010, 13.6 million tons (Mton) were produced in rain-fed areas and 14.5 in irrigation zones, the latter in a surface extension equivalent to 30 % of the rain-fed land. These three basic crops occupy 50 % of the rain-fed lands (8.1 Mha) and 40 % (2.2 Mha) of irrigation lands.

The main crops during the agricultural year 2008–2009 were corn and wheat, which accounted for 45.6 % of sown land, 23.5 % of production in tons, and 36.4 % of the production value. A little over 25 % of rain-fed and irrigation lands are devoted to crops for feeding cattle; in rain-fed land, cultivated grass (2 Mha), sorghum grain (1.3 Mha) and fodder oat (0.7 Mha). Mexico is one of the few countries where sorghum is grown on rain-fed land and alfalfa is grown in arid and semiarid areas where aquifers are overexploited; it is also a major maize consumer (Fig. 4.1), the world's fourth largest producer and a net importer of yellow corn, around five million tons annually (Sagarpa 2011) for feeding cattle. Internal production of white corn for human consumption meets demand almost entirely.

Geographic, climatic, hydrologic, and socio-demographic characteristics are reflected by an unequal distribution in agricultural production in each of the federative entities. When classified by agricultural GDP contribution to the entity's total GDP, there are three categories: those that have higher agricultural GDP to the national mean (4 %), those that are around the mean, and the entities that have an agricultural GDP lower than the mean. In Fig. 4.2, it is evident that there is no coincidence with higher precipitation volumes and, therefore, not with renewable water either, so the states with the higher relative agricultural production are

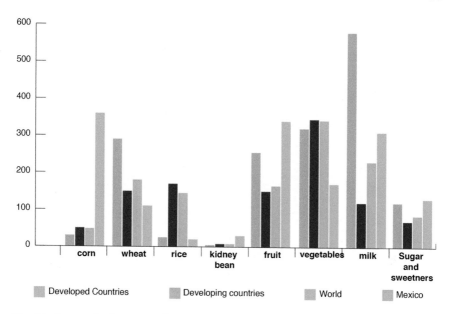

Fig. 4.1 Average foods consumption per capita. *Source* FAO (2009)

Michoacán (10 %), Nayarit (8 %), Zacatecas (7 %), Sinaloa (9 %), Durango (8 %), Oaxaca (6 %), and Sonora (6 %); on the other hand, entities with higher precipitation are those in the South and Center-Gulf areas.

Elaborating a coefficient where agricultural production cost is estimated in terms of its water use, by subtraction of the participation in agricultural GDP and the percentage of concessioned water by entity in relation to the national total, we find that only eight states contribute a higher percentage to the national agricultural production than the water they have in concession for agricultural use; these are, according to their importance: Jalisco, Veracruz, Chiapas, Oaxaca, Mexico State, Michoacán, Tabasco, and Guerrero (Fig. 4.2). The remaining federative entities take up a higher percentage of the resource in relation to the amount they contribute to the national agricultural product.

Apart from the water employed in agriculture and livestock raising, food production comprises the water used in agro-industrial transformation and commercialization processes of primary products from agriculture, and the cattle and fishing industries. In 2011, the food, alcohol, and tobacco industries accounted for 5 % of the total GDP (Sagarpa 2011). The agricultural industry as a whole contributes 22 % to the manufacturing industry; out of this percentage 17 % corresponds to the food industry and nearly 5 % to alcohol and tobacco industries (INEGI 2012).

The agro-industrial sector (processed foods, drinks, and beverages division) is formed by 12 branches: (1) meat and dairy; (2) fruits and vegetables; (3) wheat mill; (4) *nixtamal* (boiled corn) mill; (5) coffee mill; (6) sugar; (7) edible oils and fats; (8) animal food; (9) other food products; (10) alcoholic drinks; (11) malt and beer;

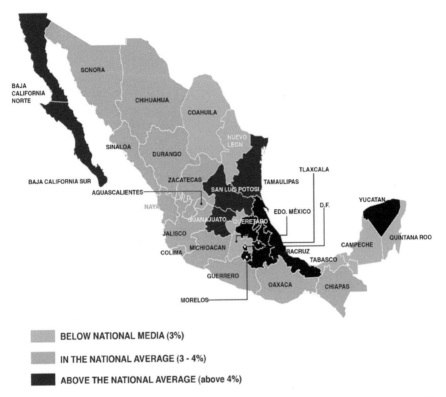

BELOW NATIONAL MEDIA (3%)

IN THE NATIONAL AVERAGE (3 - 4%)

ABOVE THE NATIONAL AVERAGE (above 4%)

Fig. 4.2 Agricultural GDP participation within state GDP, 2011. *Source* own elaboration with data from INEGI. The University Program for Metropolitan Studies, UAM, collaborated in the cartographic elaboration, with data from UNDP (2011)

and (12) soda and gasified beverages. These branches are in turn divided into 23 kinds of activities.

According to the IMTA, the food industry requires an annual volume of 435 million cubic meters (m^3) of water for its operation in the washing of raw materials, steam generation (precooking and cooking), containers washing and filling, cooling water, and equipment and floors washing. A population of six million inhabitants (200 l daily per person) could be supplied for a year with this volume.

Discharges (or gray water footprint) from the food industry (185 million m^3 of wastewaters), generate 200 thousand tons of DBO5, 151 thousand tons of total suspended solids, and 25 thousand of fats and oils annually. Other pollutants generated by the agricultural industry are fecal coliforms, chemical oxygen demand, and fats and oils (IMTA 2013). Industrial infrastructure for the food industry requires an annual supply of 214 million m^3, 33 % goes to dairy, 29 % to canned food, 18 % to bakery, 11 % to cereal mill, 3 % to edible fats and oils, 1 % to meat, and 1 % to candy and chocolate (IMTA 2013).

Taking into account water life cycle, the fact that Mexico imports almost half of its food is reflected on an external water footprint (WF) (or virtual water) from agricultural and cattle products, from a different perspective. 86 % of a Mexican individual's WF (the amount of water contained in the varied products they consume) comes from food and drink products, 6 % for other agricultural products, 5 % corresponds to domestic consumption, and 3 % to industrial products (Agroder 2012).

Cereals are the products that contain the largest amount virtual water imported by Mexico; meats and edible residue being the second most important group of imported virtual water (Arreguín 2007).

Mexico's increasing dependence on food imports has a double dimension; on the one hand, soil loss and water pollution are reduced via a lower use of agrochemicals; on the other hand, food security and sovereignty is undermined and the country becomes more vulnerable to food price increase in the international market (see Chaps. 7 and 16 in this same book).

4.2 Conclusions

- Farming and ranching activities are the most important consumers of the hydric resource in Mexico.
- The rural sector has the worst conditions of life in the country.
- Although the rain-fed sector has the largest portion of cultivated land, it does not have the highest production, neither in physical terms, nor in value.
- As in most situations in the country, there is unequal distribution of the farming and ranching activities among the federative entities, which does not correspond with the country's water availability.
- The main agricultural products are maize, wheat, and sorghum.
- The food industry is also a major water consumer.
- The importance of agricultural products results in an external water footprint.

References

* indicates internet link (URL) has not been working any longer on 8 February 2016.

AgroDer. *Huella hídrica en México en el contexto de Norteamérica*. México: WWF, AgroDer, 2012.

Arreguín-Cortes Felipe, et al. "Agua virtual en México", *Ingeniería hidráulica en México*, vol. XXII, no. 4, (2007): 121–132.

ASERCA. "La agroindustria en México". In *Boletín Comercialización 18/08*. México: ASERCA, 2008.

Carabias, Julia and Rosalva Landa. *Agua, medio ambiente y sociedad. Hacia una gestión integral de los recursos hídricos en México*, México: UNAM/COLMEX/Fundación Gonzalo Río Arronte, 2005.

CONAGUA. *Programa Nacional Hídrico 2007–2012*, México: SEMARNAT, 2008.

CONAGUA. *Estadísticas agrícolas de los distritos de riego: año agrícola 2009–2009*, México: CONAGUA, 2010.

CONAGUA. *Estadísticas del agua en México*, México: SEMARNAT, 2012.

*CONEVAL. *Pobreza 2010. Porcentaje de la población en pobreza según entidad federativa*, http://www.coneval.gob.mx/cmsconeval/rw/pages/medicion/pobreza_2010.es.do,   Accessed April 4, 2012.

FAO. *La FAO en México, más de 60 años de cooperación 1945–2009* www.fao.org.mx/ documentos/Libro_FAO.pdf, Accessed April 24, 2012.

Guerrero, Vicente. "Aportes de la gestión integral del agua". In *Hacia una gestión integral del agua en México: retos y alternativas*, edited by Cecilia Tortajada, Vicente Guerrero and Ricardo Sandoval, 31–46, Mexico: Centro del Tercer Mundo del Manejo del Agua, Mexico, Miguel Ángel Porrúa, 2004.

IMTA. *Agua para la producción de alimentos*, http://www.imta.mx/index.php?Itemid= 106&catid=52:enciclopedia-del-agua&id=180:agua-para-produccion-de-alimentos&option= com_content&view=article , Accessed October 15, 2013.

INEGI. *Perspectiva estadística*, México:INEGI, 2011.

INEGI. *El sector alimentario en México 2012*, México: INEGI, 2012.

Procuraduria Agraria. *Estadísticas agrarias*, México: Procuraduría Agraria, 2011.

SAGARPA. *Monitor agroeconómico e indicadores de la agroindustria*, http://www.sagarpa.gob. mx/agronegocios/Documents/pablo/Documentos/MonitorNacionalMacro_cierre%202011.pdf, Accessed October 16, 2013.

SEGOB. *Ley de Aguas Nacionales*, México: Diario Ofcial de la Federación, SEGOB, 2004.

Sistema de Información Agroalimentaria y Pesquera, *Agricultura. Producción agrícola,* http:// www.siap.gob.mx/index.php?option=com_content&view=article&id=42&Itemid=12, Accessed April 25, 2012.

Sistema de Información Agroalimentaria y Pesquera. *Agricultura. Producción agrícola,* http:// www.campomexicano.gob.mx/portal_siap/Integracion/EstadisticaDerivada/ComercioExterior/ Estudios/Perspectivas/maiz96-12.pdf. Accessed April 25, 2012.

Chapter 5
Water Resources Inventory and Implications of Irrigation Modernization

Eugenio Gómez-Reyes, Jaime Garatuza-Payán
and Roberto M. Constantino-Toto

Abstract This chapter presents the effects on water resources created by the introduction of an irrigation modernization policy. The impact caused by the increase of irrigation coverage tends to have negative effects on water stocks because of its technological characteristics and productive practices in the primary sector.

Keywords Water resources · Irrigation challenges

5.1 Water Resources Inventory and Infrastructure

For management and preservation purposes, since 1997 the country has been divided into 13 Hydrological-Administrative Regions (HAR), which are formed by grouping hydrographic basins. A basin is the basic unit for water resources management, but its limits consider municipalities to facilitate the integration of socioeconomic information (Conagua 2011a).

In Mexico, there is a marked contrast between population living in a given HAR, production value, and renewable water resources. HRA XIII, Mexico Valley Waters, contains almost 20 % of the total population; it contributes a little more, 2 out of every 10 pesos of the national GDP, and it is located in a region with less than 1 % of renewable water resources (Fig. 5.1).

Every year, Mexico receives approximately 1,489 billion cubic meters (km^3) in the form of precipitation. It is estimated that 73.1 % of this volume evaporates and returns to the atmosphere; 22.1 % runs off rivers and streams, and the remaining 4.8 % naturally filters through the subsoil and recharges the aquifers. Considering water exports and imports with neighboring countries and incidental recharge, annually the country has 460 km^3 of renewable fresh water.[1]

---

[1]Maximum quantity of water that can feasibly be used every year in a region; it is the quantity of water that is renewed by rainfall and the water that comes from other regions or countries (imports). It is calculated as the mean natural annual internal surface runoff, plus the total annual recharge of aquifers, plus water imports from other regions or countries, minus the water exports to other regions or countries. In the case of Mexico, the mean values are calculated from studies carried out in each region (Conagua 2011c).

© The Author(s) 2016
R.H. Pérez-Espejo et al. (eds.), *Water, Food and Welfare*,
SpringerBriefs in Environment, Security, Development and Peace 23,
DOI 10.1007/978-3-319-28824-6_5

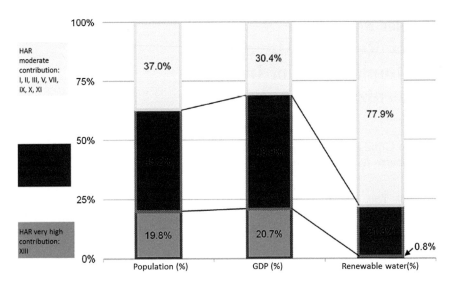

Fig. 5.1 Contrast in contribution to population, regional gross domestic product and renewable water availability, 2009. *Source* Conagua 2010a *Note* Hydrological-Administrative Regions: I Baja California Peninsula; II Northwest; III Northern Pacific; IV Balsas; V Southern Pacific; VI Rio Bravo; VII Central Basins of the North; VIII Lerma-Santiago-Pacific; IX Northern Gulf; X Central Gulf; XI Southern Border; XII Yucatan Peninsula; XIII Waters of the Valley of Mexico

According to Conagua (2011c), national precipitation is around 760 mm per year; renewable water in the country is just over 460 million cubic hectometers. In 2009, water availability per capita was 4,262 m³/inhabitant/year (Fig. 5.2).

The rivers and streams of Mexico constitute a hydrographic network of 6,33,000 km; 50 main rivers stand out, through which 87 % of the surface runoff flows. Two-thirds of the surface runoff belongs to seven rivers: Grijalva-Usumacinta, Papaloapan, Coatzacoalcos, Balsas, Panuco, Santiago, and Tonala. The surface area of their watersheds represents 22 % of the surface of Mexico (Conagua 2011c: 27).

In Mexico, evaluation of water quality is carried out using three indicators: biochemical oxygen demand (BOD), chemical oxygen demand (COD), and total suspended solids (TSS). In 2009, it was determined that 21 basins were classified as heavily polluted in one, two, or all three of these indicators. According to BOD, 13 % of surface waters is polluted; 31 % according to COD; and 7.5 % according to TSS.

As for groundwater, which is divided into 633 aquifers (Conagua 2012), it represents almost 37 % of the total volume allocated for offstream uses. From the 1970s, the number of overexploited aquifers has increased considerably, from 32 aquifers in 1975, 80 in 1985, to 105 in 2010. From these overdrafted aquifers, 54 % of groundwater is extracted for all uses, which has caused 32 aquifers to suffer the phenomenon of saltwater intrusion and/or soil salinization and brackish groundwater (Conagua 2011c: 35).

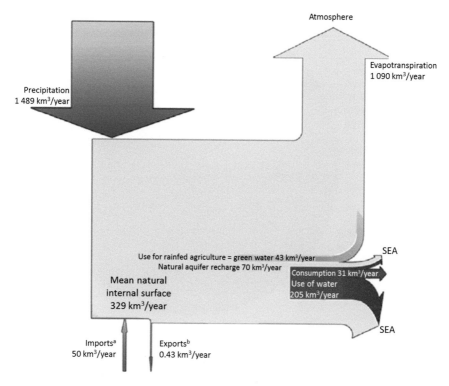

Fig. 5.2 Mean annual values of the components of the hydrologic cycle in Mexico (km^3). *Source* Conagua 2011c. *Note* The mean annual precipitation refers to the period 1971–2000. The remaining values are averages reported for 2009. The natural aquifer recharge reported in the figure, plus 9 km^3 of incidental recharge, constitute the total mean recharge

Water volume classification for offstream uses has four main groups: agriculture, public supply, self-supplying industry, and electricity generation. As shown in Fig. 5.3, the greatest water allocation is that corresponding to agricultural activities (61.8 km^3), followed by public supply (11.4 km^3), electricity (4.1 km^3), and self-supplying industry (3.3 km^3). It should be noted that 63 % of the water used in Mexico for offstream uses comes from surface water sources (rivers, streams, and lakes), whereas the remaining 37 % comes from groundwater sources (aquifers). In the period 2001–2009, surface water allocation grew by 15 %, while underground water increased by 21 %.

In 2009, hydropower plants (instream use) employed 136.1 km^3 of water to generate 26.4 TWh, which represents 11.3 % of the electricity produced in Mexico. These plants have an installed capacity of 11,383 MW, or 22 % of the country's total. At national level, the stress exerted on hydric[2] resources is moderate;

---

[2]Percentage of water used for offstream uses as compared to the renewable water resources. Indicator of the water stress in any given country.

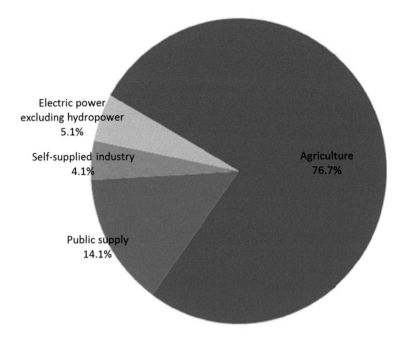

Fig. 5.3 Allocation volume distribution for offstream uses, 2009. *Source* Conagua (2009)

however, XIII Waters of the Valley of Mexico HAR is under high stress; seven HAR (Baja California Peninsula, Northwest, North Pacific, Balsas, Bravo River, North Central Basins, and Lerma-Santiago-Pacific) are under a strong degree of stress; Northern Gulf region is under low stress and four HAR (South Pacific, Central Gulf, South Border, and Yucatan Peninsula) have no stress.

Among the hydraulic infrastructure available within the country to provide the water required for the various national users, the following stands out: 4,462 dams and water retention berms; 6.50 million hectares with irrigation; 2.9 million hectares with technified rainfed infrastructure; 631 treatment purification plants in operation; 2,029 municipal wastewater treatment plants in operation (Conagua 2011c: 58).

National drinking water coverage is about 90 %, being higher in urban zones (95 %) than in rural areas (70 %). These values have increased since 1990, which were 89.4 and 51.2 %, respectively. National sanitation coverage is about 85 %, composed of 94.5 % coverage in urban areas and 57 % in rural zones; in 1990, the sanitation coverage was 79 and 18.1 %, respectively (Fig. 5.4). The greatest backlogs under both headings are in the regions Southern Pacific, Northern Gulf, Central Gulf, and Southern Border (Conagua 2011c: 68), where water is more abundant.

Urban zones annually discharge 7.49 km^3 of wastewater (municipal wastewater discharge); 6.59 km^3 (88 %) are collected in the public sewerage and 2.78 km^3

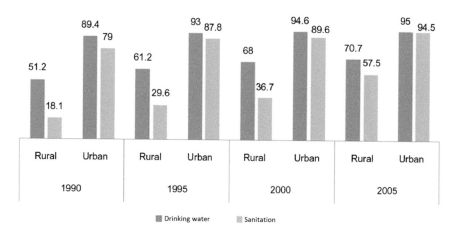

Fig. 5.4 Evolution in the rural and urban populations with drinking water and sanitation coverage in Mexico. *Source* Conagua (2011d)

(37 %) are treated. The industrial wastewater discharge is about 6 km^3/year and only 1.16 km^3 (19 %) are treated.

5.2   Irrigation Systems

In Mexico, the area with infrastructure that allows irrigation is approximately 6.5 million hectares (Mha), of which 3.5 Mha corresponds to 85 irrigation districts (Table 5.1 and Fig. 5.5) and the remaining 3 Mha to more than 39,000 irrigation units for rural development (Urderales in Spanish). Information from Conagua (2010b) notes that in the agricultural year 2008–2009, the irrigated area in the irrigation districts was 2.6 Mha (Table 5.1), similar to the historical mean of 2.5 Mha. Sinaloa, Sonora, Tamaulipas, Baja California, and Michoacan accounted for 71.1 % of the irrigated surface (Fig. 5.5).

The DR and the Urderales were designed according to the prevailing technology in the 1940s; however, even today, in 88 % of the irrigated surface of the DR water is employed by gravity, usually flooding the furrows (Fig. 5.6).

Gravity irrigation has a low total efficiency (driving efficiency plus application efficiency) in water use, between 10 and 49 %. 12 % of the irrigated area is technified with pressurized multi-floodgate, sprinkler, drip, and streak systems, with an overall average efficiency in water use between 56 and 80 % (Table 5.2).

Irrigation by sprinkler, drip, and **streak** are relatively new techniques that require a greater initial investment and a more intensive management than gravity irrigation, but they imply an important increase in the efficiency of water use, estimated in 130 % with respect to gravity irrigation.

Table 5.1 Surface irrigation use in irrigation districts by hydrological-administrative region

HAR	Name	DR No.	Area (ha)		Allocation volume (m³/s)	Irrigated area (ha)		Pumping wells	Flows
			Total	Irrigated		Gravity dams	Bypass		
I	Baja California Peninsula	2	246,906	226,041	86.711	0	129,064	96,977	0
II	Northwest	7	502,281	391,473	124.973	315,273	0	76,200	0
III	Northern Pacific	9	7,89,034	755,451	311.118	725,933	27,563	1,813	142
IV	Balsas	9	225,511	151,325	77.623	73,923	77,402	0	0
V	Southern Pacific	5	75,389	28,460	17.079	22,990	5,470	0	0
VI	Rio Bravo	12	554,597	368,433	103.409	349,229	13,505	5,699	0
VII	Central Basins of the North	1	1,16,577	69,820	32.463	69,820	0	0	0
VIII	Lerma-Santiago-Pacific	14	4,99,237	318,291	131.329	168,633	64,989	58,296	26,373
IX	Northern Gulf	13	2,65,594	135,960	48.221	86,955	27,757	0	21,248
X	Central Gulf	2	43,508	31,248	25.246	0	31,248	0	0
XI	Southern Border	4	36,399	25,970	11.219	5,152	20,445	373	0
XII	Yucatan Peninsula	2	36,871	10,051	2.095	0	0	10,051	0
XIII	Waters of Valley of Mexico	5	1,04,998	79,611	50.16	22,351	54,884	0	2,376
Total	Mexican Republic	85	3,496,902	2,592,134	1,021.65	1,840,259	4,52,327	249,409	50,139

Agricultural year 2008–2009. *Source* Conagua (2010b)

HAR Hydrological-Administrative Region. *2008–2009

DR Irrigation District (in Spanish)

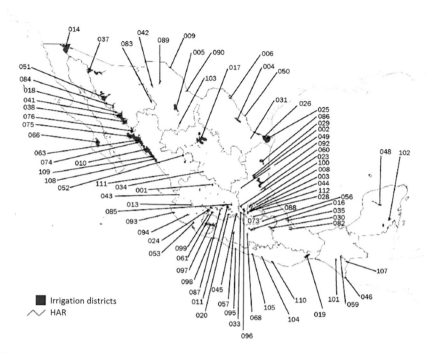

Fig. 5.5 Irrigation districts, 2009. The numbers indicate the DR no. assigned by Conagua *Source* Conagua (2011c)

The DR of the XII Yucatan Peninsula HAR are the only ones that are completely technified; however, the irrigation surface (only 10,000 ha) is the smallest of the country. In the DR of the rest of the HAR, the technified surface is less than 50 % of the total irrigation area.

The hydro-agricultural programs of the Federal Public Administration, supported by the National Water Commission (Conagua in Spanish), the Secretariat of Environment and Natural Resources (Semarnat in Spanish), and the Secretariat of Agriculture, Livestock, Rural Development, Fisheries and Food (Sagarpa in Spanish), as well as their projects and financial resources, and the modernization and rehabilitation of the irrigation districts, farm development, efficient use of water and electricity, full use of hydro-agricultural infrastructure and technified irrigation programs are important programs designed to increase production based on the most efficient use of water and hydraulic infrastructure.

However, despite the enormous resources at its disposal, public policy to modernize and technify irrigation has not achieved its main goal, which is to save water. Conversely, it has caused unwanted effects by not considering important decision-making factors (CDRRS, PEC 2007–2012).

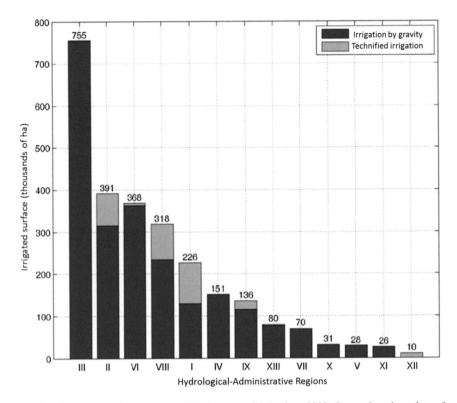

Fig. 5.6 Surface distribution in the HAR by type of irrigation, 2009. *Source* based on data of Table 5.1

Some of the problems related to irrigation policy are: the lack of irrigation user participation, except for the large hegemonic farmers; the lack of an ecosystem vision, which has not taken into account water bodies requirements; a program offering without having defined a target population; the poor training of the majority of farmers to access more complex technologies; credit shortage for participating in programs that require significant contribution from producers; operation, more formal than functional of the water institutions: Basin Councils and Organizations, Groundwater Technical Councils.

Agricultural water public policies have caused collateral problems, such as the use of an increasing water volume in irrigated areas; aquifer overexploitation because of the electricity subsidy in agricultural pumping; the lower water return due to more efficient irrigation systems (Huffaker 2010) and an increased water pollution because an increase in the use of water corresponds to a higher use of agricultural inputs, fertilizers, and pesticides.

Table 5.2 Estimated efficiencies for different irrigation systems

Irrigation system	Efficiencies (%)			Average total efficiency increment (%)[a]
	Driving	Application	Total	
By gravity	35–70	30–70	oct-49	
By sprinkler				
(a) simple	90–95	60–85	54–81	130
(b) mechanized				
1. wheel line	90–95	67–86	59–82	140
2. traveler	90–95	61–75	55–72	115
3. frontal advance	90–95	72–89	65–85	154
4. center pivot	90–95	73–90	66–86	158
Micro-irrigation				
(a) drip	90–95	85–95	77–90	183
(b) micro-sprinkler	90–95	83–93	75–88	176
(c) bubblers	90–95	80–90	72–86	168
By low pressure				
1. tubing	90–95	30–70	27–67	60
2. (a) tubing with gates (d./t)	90–95	35–72	32–69	70
(b) driv./canal	35–70	35–72	13–51	9
3. (a) tubing with gates and intermittent valve (driv. by tube)	90–95	55–85	50–81	120
(b) driv. by channel	35–70	55–85	20–60	36

[a]With respect to gravity irrigation
Source Arana-Muñoz and Monroy

References

Arana-Muñoz Omar and Oscar Monroy-Hermosillo. "Demanda de agua no potable en el Valle del Mezquital". In *Evaluación y análisis de las perspectivas para el abastecimiento de agua al Distrito Federal. Proyecto F.00396*, México: DGCOH, 2000.

CONAGUA. *Datos estadísticos,* México: CONAGUA, 2009.

CONAGUA. *Reporte económico de administración del agua*, México: CONAGUA, 2010.

CONAGUA. *Estadísticas agrícolas de los distritos de riego, Año agrícola 2009*, México: CONAGUA, 2010.

CONAGUA. *Compendio estadístico de administración del agua*, México: CONAGUA, 2011.

CONAGUA. *Gaceta de administración del agua*, México: CONAGUA, 2011.

CONAGUA. *Estadísticas del agua en México: instrumentos de gestión del agua en México*, México: CONAGUA, 2011.

CONAGUA. *Situación del Subsector Agua Potable, Alcantarillado y Saneamiento*, México: CONAGUA, 2011.

CONAGUA. *Atlas digital de México,* http://www.CONAGUA.gob.mx/atlas/ciclo21.html, Accessed April 26, 2012.

Comisión Intersecretarial para el Desarrollo Rural Sustentable. *Programa Especial Concurrente para el Desarrollo Rural*, http://www.sagarpa.gob.mx/tramitesyServicios/sms/Documents/pec2007-2012.pdf, Accessed May 3 2012.

Huffaker, Ray. "Protecting water resources in biofuels production", *Water Policy 12* (2010): 129–134.

Chapter 6
Manifestations of Welfare Loss

Úrsula Oswald Spring

Abstract Water security (WS) evolved toward the protection against floods, droughts, plagues, environmental services protection, health preservation, and conflict negotiation. The part of human and gender security deepening, as well as environmental security proposing a great (HUGE) security. This overcomes the political-military vision where water was used as a weapon. A combination of market failures, inefficient institutions, and lack of governance have aggravated conflicts and provoked violent outbreaks. Specialists have insisted on political mechanisms and negotiations between governments and those affected, which have brought about international treaties and water regimes. Growing citizen complaints about deterioration in quality and water shortage have transformed the demands for water into a basic human right, making a distinction between use value (survival) and exchange value (merchandise), with progressive tariffs for saving. Thus, WS is oriented toward people and peace, where participative governance and pacific negotiation of conflicts drive the recovery and protection of ecosystems as the lead of sociopolitical practice and an indicator of socio-environmental progress, and where science offers methodologies, methods, and proposals for norms and laws that are capable to protect the planet's future and the survival of humankind.

Keywords Water security · Loss of wellbeing · Integrated water resource management (IWRM) · Water conflicts · Hydro-diplomacy · HUGE security

6.1 Introductory Notes

Both blue and green water fostered life and ecosystem services on planet earth. Water as a transversal element is related to multiple aspects of everyday life and it becomes one of the vital elements of ecosystem services. The concept of water security (WS) has been developed as a response to these complex challenges. This chapter analyzes first the evolution of the WS concept by examining it as a political notion at local, national, regional, and international contexts. It establishes analytical relationships with human, environmental, and gender security until it becomes a 'HUGE' security that goes beyond water use and distribution. Then, it

R.H. Pérez-Espejo et al. (eds.), *Water, Food and Welfare*,
SpringerBriefs in Environment, Security, Development and Peace 23,
DOI 10.1007/978-3-319-28824-6_6

explores its relationship to economic security. After that, a case that water is also a *political and military security* issue is made. The following segment penetrates the complex political constellations that are aggravating due to climate change and the increase in hydro-meteorological events. The conclusions show that the WS concept, besides being a scientific instrument, allows the analysis of water needs and rights, from the international scope to the local community.

6.2 Evolution of the Water Security Concept

Clean water guarantees health and welfare, while polluted water may cause vector or hydric diseases. There is a close relationship between poverty, lack of water, bad quality of it, and misery, especially in rural areas and favelas, where the largest number of people in extreme poverty are found. They lack tap water, drainage, and sanitation services. WS improvement protects the population from floods, mitigates droughts, and combats plagues, particularly when the incumbent authorities collaborate with the population and generate an integral water culture.[1] While environmental and health services are closely related to WS (Ávelar et al. 2011), therefore, the consolidation of WS is not related only to military security, but also to an integral water management from the mountain to the ocean; including basin management, potable water infrastructure, sanitation and treatment, restoration of deteriorated natural environments, preservation of biodiversity, and the protection of environmental services. That is why, WS is part of a widened, deepened, and sectored security (Oswald/Brauch 2009a, b; Brauch et al. 2008, 2009, 2011), where complex interrelationships among the environmental quartet (water, air, soil, and biota) and the social one (demographic growth, urbanization, rural development, and productive processes) present themselves.

Holistic water management establishes a balance between protection, production, and use (GWP 2012). Ministers attending the Second World Water Forum in The Hague (WWF 2000) have reached a consensus for WS as

> [...] water resources and related ecosystems that provide and support the vital liquid, threatened by pollution, non-sustainable management, changes in soil use, climate change and many other forces [...] guaranteeing fresh water, coastal and related ecosystems be protected and improved to promote sustainable development and political stability, in order that every person has access to sufficient potable water and accessible prices to foster a healthy and productive life, and where vulnerable groups are protected from risks from hydro-meteorological events.

Nevertheless, in 2009 the World Health Organization (WHO) informed that nearly 900 million people did not have access to potable water and more than 2.6

---

[1]Water security is closely linked to soil security and food security, where soil fertility allows abundant harvests, stops land degradation, and combats desertification. All this affects people's welfare, reduces greenhouse gases (GHG), and improves extreme climate conditions, thanks to the vegetable cover and evapotranspiration.

million lacked sanitation services. If we add the temporary lack and scarcity of water, WS of 80 % of the world population is still being threatened. This conceptualization has allowed to understand the topic of water beyond community supply and to view its global and multidimensional aspects, wherein *virtual water*, that is, the trade of imported goods that allows to reduce pressure on available domestic resources, was included (WWC 2012). Hoekstra/Hung (2002) introduced the concept of *water footprint* to quantify all direct or indirect use of water within a nation. In turn, business people used the WS concept in a very restricted way.

6.3 Widening Water Security Toward Environmental, Social, and Economic Security

The widening of the concept of water security, developed by the Copenhagen school,[2] links WS to environmental, economic, and social security. *Environmental security* refers to the interaction of natural systems with water. Hydric systems allow the development of highly complex ecosystems with diverse ecosystem services to *provide* clean water, air, food, fibers and energetics; *regulate* the weather, purify water and air, and preserve soil fertility; *support* functions such as dissolving nutrients, disintegrating waste, and eliminating toxicity; and *facilitate the tangible and intangible cultural heritage*, where human beings integrate socially and world views reflecting society's wellbeing and peace are generated.

Three processes are deteriorating environmental security: first, a reduction in water availability due to climate change and anthropic modifications; the loss of soils' natural fertility,[3] deterioration of the vegetation cover, and the loss of biodiversity, affecting the hydric cycle. Second, degradation and pollution of water is caused by human production activities, the lack of infrastructure for the sanitation of waste domestic and industrial waters, and the diffuse pollution coming from farming and ranching activities (Pérez 2011). All these processes generate vector and hydric diseases. Third, more numerous and intense extreme

---

[2]Buzan (1997) widened the concept of military security toward environmental, economic and social security (Wæver 1995, 2008). In 1994 UNEP introduced human security as a process of deepening in security and Oswald (2008a, b, 2009a, b) added gender security to integrate equality, sustainability and equity processes in a widened and deepened vision of security. Additionally, Brauch (2008, 2009) proposed a sectorization toward water, alimentary, energetic security and so on (Oswald and Brauch 2009b; Brauch et al. 2008, 2009, 2011). All these types of security analyze the interaction between humans and nature, where the traditional hobbesian view of military and political security is surpassed.

[3]Water security is closely linked to *soil security* (Oswald and Brauch 2009b) and *alimentary security* (Oswald 2009a), where natural soil fertility permits abundant harvests, stops soil degradation and erosion, and combats desertification. All this affects people's welfare. Vegetation cover helps in fixing greenhouse gases, reducing extreme climate conditions, leaking meteorological waters into aquifers, and generating revado transpiration which moisturizes the natural environment.

hydro-meteorological events (Typhoons, floods, landslides, droughts, and forest fires) threaten life, infrastructure, and ecosystems, and generate diffuse pollution in vast areas (IPCC 2007, 2012).

Social and economic security, which complements the widening of security, is linked to human activities and welfare. Water scarcity increases social inequality, generates conflicts caused by a scarce good, and affects economic security. In Mexico, most vulnerable groups pay the highest costs for access to poor quality water, as exemplified by the Iztapalapa District in Mexico City; people there receive water sporadically and they buy it from pipetrucks and/or bottled, at prices up to 20 times higher than subsidized water that regularly reaches high-income residential areas. This not only has a negative impact on their budget and increases existing inequality, but it also becomes a permanent health risk due to the water's poor quality.

In terms of *economic security*, water value is related to three philosophical statements. 1. Plato's philosophy asserts that value is absolutely independent from things; 2. The nominal approach links a good's value to subjectivity, 3. Nominalism states that moral value is only given to a certain thing through its appreciation, which introduces ethics into the theory of value (Terricabras 1994:3637). Scheler shows in his pure axiology that validation, valuation, objectivity, quality, and hierarchy lead to negative polarities, that is, no values. Because of that, they are not independent, but they ate located within a sociopolitical system. Facing this increasing disconformities caused by water quality deterioration, ethical citizen demands came arose asking for the right to water to be enshrined as a basic human right, which has forced economics to distinguish between use and trade value of water (Ramos 2004; Oswald/Brauch 2009a). Water use value should guarantee every human being the right to access enough clean water necessary for human survival. In October 2011, the Senate of the Republic declared water a basic human right in the Mexican Political Constitution.

In the logic of trade value, economic goods are produced to optimize earnings, and supply and demand are regulated by market mechanisms. Thus, water becomes a merchandise. National and international groups have put pressure on the Mexican government to privatize water services, under the notions of making service more efficient, improving quality, and guaranteeing the sanitation of waste waters (OECD 2011). However, little investment and high rent payments to stakeholders have generated ethical problems, where poor people and immigrants in favelas do not have the resources to pay for costly services, remaining without piped water (Barkin 2011). Besides these, privatizations have released the government from its duty to fulfill the Mexican Constitution and provide every Mexican citizen with the required amount of water for their physical, social, and cultural reproduction.

Economic security proposes to establish tariffs and subsidization in order to make water available to everyone. Efficient information, administration, account-ability, crossed subsidization systems, and regional development processes should reduce hydric stress, foster water saving and reusing, as well as bind authorities and industries to guaranteeing treatment for black waters (Martí et al. 2011). Progressive tariffs lead to water savings and at the same time improve environmental, economic, and social security, since this implies higher water availability, and efficient systems may guard water quality.

6.4 Deepening Water Security in the Context of Human and Gender Security

Clean water guarantees health and welfare and, by contrast, there is an interrelation between poverty, lack of water, and bad water quality, especially in rural and urban zones in extreme poverty, where there is a lack of piped water, sewage, and sanitation. Therefore, consolidating WS does not mean to militarize water; in fact, the opposite, it is a part of a widened (environmental security) and deepened (human and gender security) security concept, since sustainable interaction between nature's needs and human and their socio-productive processes' needs is provided (Oswald 2008a). Deepening security was proposed by the United Nations Development Program (UNDP) (1994) with human security, and it has received pressure from multiple fronts in subsequent years. It was first understood as *an absence of fear*, from a Canadian approach; then as *an absence of needs* in order to generate quality of life, health, education, and welfare (CHS 2012); third, as the *absence of disasters* in the face of extreme meteorological events (Bogardi/Brauch 2005), as set forth by UN-EHS, where vulnerable population exposed to external events is protected; and finally, as the *freedom to live with dignity*, Kofi Annan (2005) championed the rule of law, water as a human right, and the pacific conciliation of water conflicts. This deepening was complemented with gender security (Oswald 2008a, b; 2009a, b; Serrano 2009; Table 6.1), where the referent object changed from the State to vulnerable groups; the values at risk are not territorial sovereignty, but gender relationships, discrimination, equity, and identity. From this perspective, threats do not come from other States, but from the patriarchal system, marked by violent, excluding, authoritarian and domination relationships, exercised by elites, authoritarian governments, and churches, and become more acute within families on account of intolerance.

6.5 Water Security as a Political and Military Security Issue

In spite of conceptual innovations, political and military security is still related to WS, when opposing national and international policies between States and non-State actors (multinational companies) have created conflicts. Water has historically been used as a weapon (floods, well poisonings, besieged cities, and destruction of water infrastructure, such as the one sustained in the Yugoslavian and Iraqi conflicts), or a bargaining mechanism (Jordan, Nile, Bravo river basins, and so on). Limited or shared resources have led to international treaties (Wolf 1999;

Table 6.1 Human, gender, and environmental security: a great (HUGE) security

Expansion level	Determination which security?	Expansion mode, reference object security for whom?	Values at risk security of what?	Threats sources security from who or what?
No expansion	National security (political, military dimension)	The state	Sovereignty, territorial integrity	Other states, terrorism, sub-State actors, guerrilla
Increased	Societal security	Nations, vulnerable societal groups	National unity and identity	(States), nations, immigrants, foreign cultures
Radical	Human security	Individuals (human kind)	Survival, quality of life, cultural integrity	The State, globalization, nature, climate change, poverty, fundamentalism
Ultra-radical	Environmental security	Urban and agricultural ecosystem	Sustainability	Nature, human kind
Trans-radical	Gender security	Women, children, indigenous, elderly, minorities	Gender relationships, equity, identity, social relationships	Patriarchy, totalitarian institutions (elites, governments, religions, culture) intolerance

Source Møller (2003); Oswald/Sandoval (2005)

1999) and water regimes (Kipping 2009; Borghese 2009; Lindeman 2009). Gleick (1993, 1998, 2001) argues, analyzing 3500 years of water conflicts, that a few times water is the only factor to unleash armed conflict and usually complex political-environmental constellations determine violence. Today, a combination of market failures, institutions incapable of managing water, and the lack of governance have aggravated conflicts and caused violent episodes.

In the face of this risk, military security specialists (Fig. 6.1) have insisted on political mechanisms and the negotiating capacity between governments and those affected in regions with hydric stress and a lack of money to create infrastructure. By integrating human, gender, and environmental security (a 'great' HUGE security; Oswald 2009a, b), negotiating for the vital liquid is promoted, technical solutions are proposed, as well as integral management for all the basin which prioritizes efficiency, equity, and sustainability in hydric and hydraulic management, thanks to technical and administrative methods.

6.6 Conflicts and Hydro-Diplomacy

Hydric conditions in Mexico are adverse and the global environmental change is aggravating them. In regions with high population and production concentrations there are seasonal—water only in rain seasons—spatial—arid and semiarid conditions—and structural—megalopolis—which have limited supply. 77 % of the total population live in zones that have 31 % of water availability, and 87 % of the GDP production takes place in those areas as well (Conagua 2008, 2009). Precisely in this region, 104 of the 653 aquifers are overexploited (Arreguín et al. 2011). For instance, the seven most overexploited aquifers in the world are located in the Mexico Valley and climate change has reduced precipitation (Rosengaus 2007). The authority is facing a dilemma: a larger demand, where there is no water and, therefore, an increasing hydric stress; or bringing in water from other basins, thus affecting other regions.

In the face of this dilemma, Conagua identified the most outstanding conflicts, analyzed involved interests, and established the steps for negotiation. The *National Water Law* (LAN, 2004) defines the priority order for water use: human consumption, services, industrial needs, agricultural use, and green water for ecosystems. This preference order affects environmental services which are precisely the water suppliers and, therefore, hydric stress keeps growing.

Among the many water conflicts in the country is the case of the 13 Towns of Morelos, where the Chihuahuita Spring, burdened with scarcity, should supply thousands of small social houses and apartment complexes. Mobilization forced the State's government and private investors to generate alternative supply sources and cancel some of the housing projects. The Lerma–Cutzamala system supplies the city of Toluca and the Mexico Valley (Perló/Gonzáles 2009; Oswald 2005) and indigenous women of the Zapatista Army and Mazahua women for the Defense of Water, armed with farming tools and wooden rifles, occupied the purification plant that provides 19 m^3/s to the metropolitan areas. Slow negotiation has made matters worse and the Mazahua people are still living in extreme poverty and with little access to water and sanitation. Another conflict unfolded in 2012 due to the construction of an aqueduct intended to run from the El Novillo dam to the coast of Hermosillo, leaving indigenous population and farmers of the Yaqui Valley without water, as 77 % of the watering surface in an arid and semiarid surrounding was sacrificed for the benefit of potential voters in the capital city of Hermosillo.

In order to negotiate these conflicts peacefully we propose a model of hydro-diplomacy, which establishes causal relationships between supply (the environmental quartet: water, air, land, biodiversity) and demand, in spite of growing populations, urbanization, industrialization, and new services. By controlling leaking and waste, piped water availability can be increased up to 70 % (Fig. 6.1). Through the restoration of ecosystems, rainwater harvesting, and integral basin management (IWRM, Jansky et al. 2008), improving water quality and quantity, the causes for conflict are reduced. Sanitation and potable water operating systems are grouped by the National Association of Water and Sanitation Companies (ANEAS). Farmers, who use 78 % of the total water, formed the National Association of Watering Units (ANUR) and businessmen organized in the Enterprise Council and the government should, as a referee, guarantee the basic human right to water to every citizen. Integral negotiation should satisfy all parties

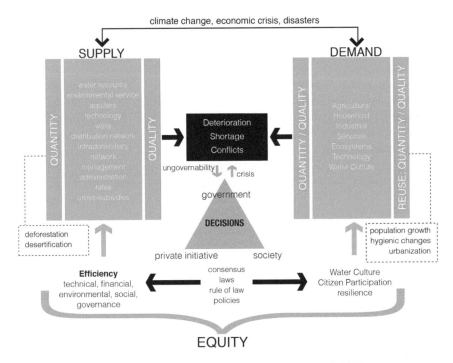

Fig. 6.1 Water security and social consensus. *Source* Oswald/Sandoval (2005)

involved in the conflict and be complemented with public policy and infrastructure. This would lead Mexico to a new water culture with the conciliation of opposed interests, more efficiency, equity, and sustainability.

6.7 Water Security and Gender Equity

Water management with WS is highly complex and requires an integral water culture that offers all social groups a less uncertain future and one more environmentally, socially, politically, and culturally sustainable. Women are a particularly vulnerable group, but at the same time, crucial in consolidating water culture. They handle water at home and when the need for sanitation is neglected, they usually take care of those affected by diarrheas and vectors (Dengue fever). The government has the responsibility to offer water free of harmful organisms (bacteria, viruses, and protozoa), organic and inorganic toxic substances, and acceptable in terms of color, odor, and taste, but within the family context, women, who carry water in times of scarcity, incorporate sanitation and hygiene to their daily activities, as well as frequently supplying their families—in the rural context—with food from their garden.

The subject of toxic metal-related diseases is particularly delicate. Arsenic, fluor, cadmium, and iron are known to generate cancer, which affect the immune system and the intellectual development of children. In Mexico, it is estimated that at least 400 thousand people have severe damage due to arsenic (INE 2012) and in cities like Aguascalientes, there is 18 % of children with kidney infections (Arreola et al. 2011; Avelar et al. 2011). As far as death by diarrhea, 212.3 out of every 100 thousand children under 5 died in 1984, while this figure fell to 60.4 in 1993, although 80 % of diseases in young children and 50 % of cases of infant mortality are associated with polluted water.

Climate change has increased vector-related diseases: malaria increased from 2.77 to 7.27 cases/100 thousand people/year between 2000 and 2005 and it is estimated that 30 % of the population is at risk, while dengue fever is expanding even more rapidly. States in the central region of the country, such as Morelos, had an increase of 600 % in 2010, and dengue fever threatens 80 % of people in the south and east regions (DGE-SSA 1984–2011). This situation could become worse because of climate change (IPCC 2007, 2012).

Deaths during disasters also show gender differences: 63 to 78 % of those killed in the great Tsunami in Asia, 80 % in the Pakistan earthquake (Ariyabandu/Fonseka 2009), and 72 % in hurricane Stan were women and girls. These deaths are related to poverty, discrimination, lack of training, and social marginalization (Oswald 2011a, b; 2012a). Additionally, there is an intervention of social representations created within every society (Serrano 2010), where women were socialized and assumed the role of caregivers and save others even at the expense of their own lives during times of disaster. Poor women, heads of households, are the most vulnerable. Poverty has a female face and 70 % of the 1,200 people who live in extreme poverty in the world are women (UNDP 2011). In times of disaster, government helps privileged male heads of family and there is institutional discrimination in aid distribution. Reducing gender vulnerability in the face of climate change and disasters depends on national and regional public policy with concrete practice that fosters equality and equity in all life spheres, production and political activities, and creates resilience.[4]

6.8 Concluding Notes: Toward a Comprehensive Understanding of Water Security

Facing these complex and contradicting processes, the notion proposed is to integrate WS to human, gender, and environmental security, and to promote a 'great' or HUGE security (Oswald 2009a, b, 2012a, b). HUGE points human security toward people and peace challenges, gender security toward equity, and environmental

---

[4]Resilience refers to a process capable to anticipate adverse and complex natural and social phenomena, through planning and learning from previous disasters and the capacity to respond in a flexible way in the presence of unknown processes and threats, in order to reduce social vulnerability, protect vulnerable people, and rapidly recover socially after an extreme event.

security toward sustainability; and the subjacent structures of violence, unequal appropriation of resources, and vertical and authoritarian power structures are analyzed with it as a perspective. HUGE proposes consolidating participative democracy with governance and conflict negotiation, where violent conflicts prevention and their peaceful resolution converge with solidarity toward the vulnerable, in order to reach sustainable, diverse, and equitable developments, without institutional and social discrimination.

In terms of environmental security, it is necessary to encourage the recovery and protection of ecosystems and their environmental services, where humans represent only one species within the complex biotic interrelations, and where sustainable management should avoid that humans inflict further damage to nature. Conventional use of water in Mexico has reached its limits and nonconventional sources such as reuse, desalination, water market, virtual water, and water footprint reduction; these elements offer the government and society new tools for improving water security in the national and local contexts. The challenge for the twenty-first century in Mexico is to reach this WS by establishing a fair and sustainable balance between supply and demand. Thus, WS is the lead for new sociopolitical practice, but also an indicator of socio-environmental progress of a conscious society, capable to change. Weak organizations and unemployment may be overcome with the support of this society, and gender violence, illiteracy, lack of solidarity, envy, hunger, malnutrition, and violent conflicts can be fought with it, too. More severe and numerous natural disasters help encourage resilience from the bottom up, which will be supported by government prevention, early alert, and preventive evacuation policies, in order to limit human and material damages. Confronted by a growing water scarcity, society, honest business people, and authorities should consolidate negotiation to guarantee fair access to water and the peaceful resolution of conflicts with tolerance and equity, where social welfare and environmental recovery are prioritized.[5]

In sum, the concept of water security helps to make the transformations of limited military security transformations to a widened and deepened security explicit. Its usefulness lies on the understanding that national, regional, and international agreements, sustainable management of ecosystems, green agriculture, urban services, and gender perspective policy will allow to improve human and environmental health. Besides, integral WS becomes also a scientific task, where

---

[5]In the municipalities of Mazapil and Villa de Cos, in Zacatecas, the Canadian multinational company Goldcorp extracts water from 30 wells for washing mineral concentrations and 5,400 liters of water are polluted per gold ounce (28.3495 g). This water could be destined for human, animal, or environmental consumption to consolidate regional development. In 2012, there were175 thousand families in the region who supported themselves from agricultural activities, and could not satisfy their basic needs due to the lack of water. Droughts in the region have become more acute with climate change and life conditions have forced many people to migrate facing a survival dilemma.

methodologies, analysis methods, integral basin management, and norms and laws development are key elements for the planet's future and the survival of the human species.

References

* indicates internet link (URL) has not been working any longer on 8 February 2016.

Annan, Kofi. *In Larger Freedom: Towards Security, Development and Human Rights for All, Report of the Secretary General for Decision by Heads of State and Government,* (Nueva York, Department of Public Information, 21 de marzo de 2005).

Ariyabandu, Madhavi Malalgoda and Dilrukshi Fonseka. "Do Disasters Discriminate? A Human Security Analysis of the impact of the Tsunami in India, Sri Lanka and of the Kashmir Earthquake in Pakistan". In *Facing Global Environmental Change: Environmental, Human, Energy, Food, Health and Water Security Concepts* edited by H.G. Brauch et al., 1215–1226, Berlin: Springer-Verlag, 2009.

Arreguín, Felipe, Mario López Pérez and Humberto Marengo Mogollón. "Los retos del agua en México en el siglo XXI". In *Retos de la investigación del agua en México,* coordinated by Ursula Oswald Spring, 19–34, México: CRIM-UNAM, Conacyt, 2011.

Arreola, Laura et al. "Potable Water Pollution with Heavy Metals, Arsenic and Fluorides and its Relation to the Development of Kidney Chronic Illness in the Infant Population of Aguascalientes". In *Water Research in Mexico. Scarcity, Degradation, Stress, Conflicts, Management, and Policy,* edited by Ursula Oswald, 231–238, Berlin: Springer, 2011.

Avelar, Francisco et al. "Water quality in the State of Aguascalientes and its effects in the population's health". In *Water Research in Mexico. Scarcity, Degradation, Stress, Conflicts, Management, and Policy,* edited by Ursula Oswald, 217–230, Berlin: Springer Verlag, 2011.

Bächler, Günther et al. *Kriegsursache Umweltzerstörung. Ökologische Konflikte in der Dritten Welt und Wege ihrer friedlichen Bearbeitung,* vol. I, Chur-Zürich: Rüegger, 1996.

Bächler, Günther and Kurt R. Spillmann. *Kriegsursache Umweltzerstörung. Länderstudien von externen Experten. Environmental Degradation as a Cause of War. Country Studies of External Experts,* vol. III, Chur-Zürich: Rüegger, 1996.

Bächler, Günther and Kurt Spillmann. *Kriegsursache Umweltzerstörung. Regional- und Länderstudien von Projektmitarbeitern. Environmental Degradation as a Cause of War. Regional and Country Studies of Research Fellows,* vol. II, Chur-Zürich: Rüegger, 1996.

Bächler, Günther, Kurt Spillmann and Mohamed Suliman. *Transformation of Resource Conflicts: Approach and Instruments,* Berlin: Peter Lang Bern, 2002.

Barkin, David. "La ingobernabilidad en la gestión del agua urbana en México". In *Retos de la investigación del agua en México,* coordinated by Ursula Oswald, 539–552, Mexico: CRIM-UNAM, Conacyt, 2011.

Bogardi, Janos and Hans Günter Brauch. "Global Environmental Change: A Challenge for Human Security – Defining and conceptualising the environmental dimension of human security". In *UNEO – Towards an International Environment Organization – Approaches to a sustainable reform of global environmental governance* edited by Andreas Rechkemmer, 85–109, Baden-Baden: Nomos, 2005.

Borghese, Maëlis. "The Centrality of Water Regime Formation for Water Security in West Africa: An Analysis of the Volta Basin". In *Facing Global Environmental Change: Environmental, Human, Energy, Food, Health and Water Security Concepts,* edited by Hans Günter Brauch et al., 685–698, Berlin: Springer-Verlag, 2009.

Brauch, Hans et al. *Globalization and Environmental Challenges: Reconceptualizing Security in the 21st Century,* Berlin: Springer-Verlag, 2008.

Brauch, Hans. "Introduction: Globalization and Environmental Challenges: Reconceptualizing Security". In *Globalization and Environmental Challenges: Reconceptualizing Security in the 21st Century*, edited by Hans Günter Brauch, 27–43, Berlin: Springer-Verlag, 2008.

Brauch, Hans. "Introduction: Facing Global Environmental Change and Sectorialization of Security". In *Facing Global Environmental Change: Environmental, Human, Energy, Food, Health and Water Security Concepts*, edited by Hans Günter Brauch et al., 27–44, Berlin: Springer-Verlag, 2009.

Brauch, Hans et al. *Facing Global Environmental Change: Environmental, Human, Energy, Food, Health and Water Security Concepts*, Berlin: Springer-Verlag, 2009.

Brauch, Hans et al. *Coping with Global Environmental Change, Disasters and Security. Threats, Challenges, Vulnerabilities and Risks*, Berlin: Springer-Verlag, 2011.

Buzan, Barry, Ole Wæver, Jaap de Wilde. *Security. A New Framework for Analysis*, Boulder: Lynne Rienner, 1997.

CHS. *Human Security Now, Protecting and empowering people, Commission on Human Security*, http://www.humansecurity-chs.org/finalreport/, Accessed May 16, 2012.

CONAGUA. *Datos estadísticos*, México: CONAGUA, 2009.

CONAGUA. *Programa Nacional Hídrico 2007–2012*, México: SEMARNAT, 2008.

Conca, Ken and Geoffrey Dabelko. *Environmental Peacemaking*, Baltimore: Johns Hopkins University Press, Woodrow Wilson Center Press, 2002.

Cortés, Juana and Cesar Calderón. "Potable water use from aquifers connected to irrigation of residual water". In *Water Research in Mexico. Scarcity, Degradation, Stress, Conflicts, Management, and Policy* edited by Ursula Oswald, 189–200, Berlin: Springer Verlag, 2011.

Gleick, Peter. "Water and Conflict: Fresh Water Resources and International Security", *International Security* 18:1 (Summer), (1993): 79–112.

Gleick, Peter. "Water in crisis: Paths to sustainable water use", *Ecological Applications* 8:3, (1998): 571–579.

Gleick, Peter. "Global Water: Threats and Challenges Facing the United States. Issues for the New U.S. Administration", *Environment* 43:2, (2001): 18–26.

Gleick, Peter. *Water Conflict Chronology*, http://www.worldwater.org/conflictchronology.pdf, Accessed August 1, 2012.

*GWP, *Towards Water Security: A Framework for Action*, http://www.gwpsudamerica.org/docs/publicacoes/doc_78_en.pdf, Accessed March 4, 2012.

Hansen, Anne and Carlos Corzo. "Evaluation of pollution in hydrological basins: priorities and necessities". In *Water Research in Mexico. Scarcity, Degradation, Stress, Conflicts, Management, and Policy* edited by Ursula Oswald, 201–230, Berlin: Springer Verlag, 2011.

Hoekstra, Arjen and Pham Hung. "Virtual Water Trade: A Quantification of Virtual Water Flows Between Nations in Relation to International Crop Trade", *Proceedings of the International Expert Meeting on Virtual Water Trade*, Value of Water Research Report Series 12, (2002): 25–47.

Homer-Dixon, Thomas. "On the Threshold: Environmental Changes as Causes of Acute Conflict", *International Security* 16:2 (Fall), (1991): 76–116.

Homer-Dixon, Thomas. "Environmental Scarcities and Violent Conflict: Evidence From Cases", *International Security* 19:1 (Summer), (1994): 5–40.

Homer-Dixon, Thomas. *Environment, Scarcity, and Violence*, USA: Princeton University Press, 1999.

*INE. *Programa de monitoreo y evaluación de sustancias tóxicas, persistentes y bioacumulables*. http://wwwsiscop.ine.gob.ms/proname.html, Accessed May 26, 2012.

IPCC. *Climate Change 2007.Impacts, Adaptation and Vulnerability*, http://www.ipcc-wg2.org/, Accessed January 30, 2012.

IPCC. *Managing the Risks of Extreme Events and Disasters to Advance Climate Change Adaptation*, Cambridge: Cambridge University Press, 2012.

Jansky, Libor, Mikiyasu Nakayama and Pachova I. Nevelina. "Introduction: From domestic to international water security". In *International Water Security. Domestic Threats and*

Opportunities edited by Jansky, Libor, Mikiyasu Nakayama, Pachova I. Nevelina, 1–5, Tokyo: United Nations University Press, 2008.

Kipping, Martin. "Can 'Integrated Water Resources Management' Silence Malthusian Concerns? The Case of Central Asia". In *Facing Global Environmental Change: Environmental, Human, Energy, Food, Health and Water Security Concepts*, edited by Hans Günter Brauch et al., 711–723, Berlin: Springer-Verlag, 2009.

Lindemann, Stefan. "Success and Failure in International River Basin Management – The Case of Southern Africa". In *Facing Global Environmental Change: Environmental, Human, Energy, Food, Health and Water Security Concepts,* edited by Hans Günter Brauch, 699–710, Berlin: Springer Verlag, 2009.

Martín, Alejandra et al. "Assessment of a Water Utility Agency: A Multidisciplinary Approach". In *Water Research in Mexico. Scarcity, Degradation, Stress, Conflicts, Management, and Policy* edited by Ursula Oswald, 421–434, Berlin: Springer-Verlag, 2011.

Moller, Bjorn. "National, Societal and Human Security: Dicussion-A Case Study of the Israeli-Palestine Conflict". In *Security and Environment in the Mediterranean. Conceptualising Security and Environmental Conflicts,* edited by Hans Günter Brauch et al., 277–288, Berlin: Springer, 2003.

OECD. *Environmental Indicators-Towards Sustainable Development*, Paris: OCDE, 2011.

Oswald, Úrsula. "Sustainable Development with Peace Building and Human Security". In *Our Fragile World. Challenges and Opportunities for Sustainable Development* edited by Mostafa K. Tolba, 873–916, Oxford: Oxford-EOLSS Publisher, 2001.

Oswald, Úrsula. *El valor del agua. Una visión socioeconómica de un conflicto ambiental,* Tlaxcala: Coltlax/Gobiernos del Estado de Tlaxcala/Fondo Mixto Conacyt, 2005.

Oswald, Úrsula. "Gender and Disasters. Human, Gender and Environmental Security: A HUGE Challenge", Source 8, Bonn: UNU-EHS, 2008.

Oswald, Úrsula. "Globalization from Below: Social Movements and Altermundism. Reconceptualizing Security form a Latin American Perspective". In *Globalization and Environmental Challenges: Reconceptualizing Security in the 21st Century*, edited by Hans Günter Brauch et al., 379–402, Berlin: Springer-Verlag, 2008.

Oswald, Úrsula. "A HUGE Gender Security Approach: Towards Human, Gender, and Environmental Security". In *Facing Global Environmental Change: Environmental, Human, Energy, Food, Health and Water Security Concepts,* edited by Hans Günter Brauch et al., 1165–1190, Berlin: Springer-Verlag, 2009.

Oswald, Úrsula. "Food as a new human and livelihood security issue". In *Facing Global Environmental Change: Environmental, Human, Energy, Food, Health and Water Security Concepts*, edited by Hans Günter Brauch et al., 471–500, Berlin: Springer-Verlag, 2009.

Oswald, Úrsula. "Reconceptualizar la seguridad ante los riesgos del cambio climático". In *Las dimensiones sociales del cambio climático: un panorama desde México. ¿Cambio social o crisis ambiental?* coordinated by Simone Lucatello and Daniel Rodríguez, 23–48, México: Instituto de Investigaciones Dr. José María Luis Mora/UNAM, 2011.

Oswald, Úrsula. "Social Vulnerability, Discrimination, and Resilience-building in Disaster Risk Reduction". In *Coping with Global Environmental Change, Disasters and Security –Threats, Challenges, Vulnerabilities and Risks*, edited by Hans Günter Brauch et al., 1169–1188, Berlin: Springer-Verlag, 2011.

Oswald, Úrsula. "Environmentally-Forced Migration in Rural Areas: Security Risks and Threats in Mexico". In *Climate Change, Human Security and Violent Conflict. Challenges for Societal Stability*, edited by Jürgen Scheffran et al., 315–350, Berlin: Springer-Verlag, 2012.

Oswald, Úrsula. "Reducción de riesgos por desastres y resiliencia social: superando la discriminación, la vulnerabilidad social y los desastres en el mundo y México", *Memorias del 2° Congreso Internacional de la Society of Risk Assessment*, Bogotá: Universidad de los Andes, 2012.

Oswald, Úrsula and Hans Günter Brauch. *Reconceptualizar la seguridad en el siglo XXI*, México: CRIM/CCA/CEIICH-UNAM, Senado de la República, 2009.

Oswald, Úrsula and Hans Günter Brauch. "Securitizing Water". In *Facing Global Environmental Change: Environmental, Human, Energy, Food, Health and Water Security Concepts*, edited by Hans Günter Brauch et al., 175–202, Berlin: Springer-Verlag, 2009.

*Oswald, Úrsula and Hans Günter Brauch. *Securitizar la tierra. Aterrizar la seguridad*, http://www.unccd.int/knowledge/docs/dldd_sp.pdf, Accessed June 16, 2012. (Bonn: UNCCD, 2009).

Oswald, Úrsula and Francisco Sandoval. "Presentación Oral", *Primer Foro Universitario del Agua*, Cocoyoc: UNAM, 2005.

Pérez, Rosario. "Contaminación del agua por la agricultura: retos de política". In *Retos de la investigación del agua en México* coordinated by Ursula Oswald, 605–616, México: CRIM-UNAM, Conacyt, 2011.

Perló, Manuel and Arsenio González. *Guerra por el agua en el Valle de México*, México: PUEC-UNAM, 2009.

PNUD-UNEP. *Reporte sobre seguridad humana*, Oxford: PNUD, Oxford University Press, 1994.

PNUD-UNDP. *Informe sobre desarrollo humano. Sostenibilidad y equidad: un mejor futuro para todos*, New York: UNDP, 2011.

Ramos, Sergio. *Mercados de agua*, México: IMTA, 2004.

Rosengaus, Michael. *Informe interno. Procedimientos para estimar tendencias del análisis parcial de datos históricos, 40 años de datos diarios, Tmax, Tmin y precipitación, tendencias, promedio nacional, todos los meses y anuales de 1961 a 2000 a nivel nacional, regional y estatal*, México: CSMN, 2007.

SEGOB. *Ley de Aguas Nacionales*, México: Diario Oficial de la Federación, SEGOB, 2004.

Serrano, Erendira. "The impossibility of Securitizing Gender vis a vis Engendering Security". In *Facing Global Environmental Change: Environmental, Human, Energy, Food, Health and Water Security Concepts*, edited by Hans Günter Brauch et al., 1151–1164, Berlin: Springer-Verlag, 2009.

Serrano, Erendira. "La construcción social y cultural de la maternidad en San Martín Tilcajete, Oaxaca", PhD diss. UNAM, 2010.

Terricabras, Josep. "Teoría del valor". In *Diccionario de filosofía*, José Ferrater and Josep Terricabras, Barcelona: Ariel, 1994.

The Royal Academy of Engeneering. *Global Water Security –an engineering perspective*, Londres: The Royal Academy of Engineering, 2010.

Wæver, Ole. "Securitization and Desecuritization". In *On Security* edited by Ronnie D. Lipschutz, 46–86, New York: Columbia University Press, 1995.

Wæver, Ole. "Peace and Security: Two Evolving Concepts and their Changing Relationship". In *Globalization and Environmental Challenges: Reconceptualizing Security in the 21st Century*, edited by Hans Günter Brauch et al., 99–111, Springer-Verlag, 2008.

WMO. *Declaración de Dublín sobre el Agua y el Desarrollo Sostenible* http://www.wmo.int/pages/prog/hwrp/documents/espanol/icwedecs.html, Accessed May 16, 2012.

Wolf, Aaron. *Conflict and cooperation along international waterways*, http://www.transboundarywaters.orst.edu/publications/ conflict_ coop/, Accessed May 6, 2012.

Wolf, Aaron. "'Water Wars' and Water Reality: Conflict and Cooperation along International Waterways". In *Environmental Change, Adaption, and Security* edited by Steve Lonergan, 251–265, Dordrecht: Kluwer, 1999.

WWC, *Virtual Water Trade –Conscious Choices*, http://www.worldwatercouncil.org/fileadmin/wwc/Library/Publications_and_reports/virtual_water_final_synthesis.pdf, Accessed May 16, 2012.

World Water Forum in The Hague, Ministers at the Second World Water Forum in The Hague. *Closing Statement*, The Hague, 2000.

Chapter 7
Prices and Water: A Strategy with Limited Effectiveness

Roberto M. Constantino-Toto

Abstract This paper studies the characteristics of water markets in Mexico under the analysis of water and food security. Water management functions are related to the mechanisms of water allocation for the different areas of use: food production, environmental services preservation, social welfare promotion by ensuring direct water consumption to citizens, and economic prosperity by its productive use. Regular and stable water supply of sufficient quality, wastewater collection and treatment, and billing and collection of water consumption correspond to the Operating Organizations of Drinking Water, Sewage and Drainage. Water is a good with no substitutes, it is transversal and it constitutes a natural monopoly. The phenomenon of institutional evolution which results in a scenario of water resources management system fragility, characterized by the lack of coordination between different management levels related to water policy, is analyzed in the first part of this chapter. The phenomenon of financial sustainability required for water and food security promotion is discussed in the second part.

Keywords Water · Prices · Markets

7.1 Introduction

It is considered appropriate to establish a set of elements to clarify the meaning of the 'market' and 'price' terms utilized in this text to study the characteristics of water markets in Mexico under the analysis of water and food security.

As stated by Van der Zaag/Savenije (2006), there is confusion in the analyses that compare the functions of the entities responsible for the provision of drinking water, sewerage and drainage services, with those concerning to institutional functions related to the water management model. In this regard, it is worth noting that the functions of water management are related to the mechanisms of water allocation for the different areas of use: food production, environmental services preservation, social welfare promotion by ensuring direct water consumption to citizens and economic prosperity by using it in the productive sectors. The functions that correspond to the Operating Organizations of Drinking Water, Sewage and Drainage

© The Author(s) 2016
R.H. Pérez-Espejo et al. (eds.), *Water, Food and Welfare*,
SpringerBriefs in Environment, Security, Development and Peace 23,
DOI 10.1007/978-3-319-28824-6_7

(OOAPAS in Spanish), which are responsible for the provision of public services at local level, are regular and stable water supply of sufficient quality, wastewater collection and treatment, as well as the billing and collection of water consumption.

Something similar happens to water markets and prices. Markets often refer transaction structures for private goods, i.e., goods that possess the attributes of exclusivity and rivalry, as defined in economic theory. Extending this definition to the case of water, without the corresponding clarification, would lead to misinterpretation considering water as a private good and its social allocation process would proceed through pricing arrangements. In the case of the latter category, economic theory conventionally states that prices are the expression of the exchange value of the goods that are object of the transaction, and its formation takes place based on production costs structure and acceptable profit margins for producers. However, prices, beyond their formation, can be interpreted as carriers of information about the intrinsic difficulty for the production of goods or their availability, thereby facilitating individual allocation mechanisms.

This allows to establish the meaning of the concepts of 'markets' and 'prices' used in this document. 'Markets' refer to the set of social spaces for interaction and exchange bounded by rules and social, cultural, historical, and legal values, and that may include not only private but also public goods (those without some or none of the characteristics of private goods). 'Prices' refer to vehicles of information and mechanisms to facilitate the assignment, whose magnitude may include, not only reference to costs, but also elements related to some sort of institutional intervention in terms of constraints that inhibit adverse effects in the collective welfare. In the latter case and to avoid confusion, hereafter the author refers to water rates to highlight that this element is not just a simple good for economy or for a commodity, but one for which there are no substitutes, it is transversal and it constitutes a natural monopoly.

The characterization of water markets in Mexico and the current tariff structure are analyzed from the perspective of the possibility of institutional competencies accumulation that facilitate a transition that strengthens the management model from the perspective of water and food security.

The study is divided in two sections. In the first one, the institutional evolution that has resulted in a scenario of water resources management system fragility, whose most significant feature is the lack of coordination between different management levels related to water policy, is discussed. The second part refers to the financial sustainability required for water and food security promotion.

7.1.1   Contemporary Water Markets in Mexico: Their Institutional Origin

The current situation of the Mexican water sector is the result of a historical, institutional, and cultural evolution that has facilitated the consolidation of contemporary careless practices of use and procurement.

Collado (2008) carried out a reconstruction of how the institutional water treatment has historically evolved in Mexico, from a perspective where several significant facts stand out. First, the tradition of water use and its management as a common good related to local practices begins to fracture in the late nineteenth century. Second, the federalization of water management is associated, at the time, with the possibility of granting concessions (Aboites 2004). Third, the emergence of a definitive water management system that recognizes the existence of national waters and favors the figure of the federation in this matter, comes from the enactment of the Water Use Act of Federal Jurisdiction (1910), the Constitution (1917) and the Law on Irrigation with Federal Waters (1926).

Water legal changes have been successive and periodic (Dau 2008), and have required, in turn, changes in the management structure. However, water was interpreted at the time as a promotional vehicle. The corresponding laws, from the late 1920s to mid-1940, empowered the Ministry of Agriculture and Development (SAF in Spanish) to attend the authorization of projects and uses of water.

Population growth rate in the country, particularly in urban areas, and economic growth process supported federal participation functions in the Local Boards of Water Users, to change from a passive role to another of increasing directionality, as the investment funds required for the development of infrastructure for water supply, could only be authorized by federal representatives (Pineda 2008).

The SAF was restructured in 1946, leading to the creation of two new entities, the Ministry of Agriculture and Livestock (SAG in Spanish), responsible for water in irrigation districts, and the Ministry of Water Resources (SRH in Spanish), responsible for planning, monitoring, implementation of the drinking water and sanitation projects, when such infrastructure was financed with federal funds. In this case, the federal agencies remained at local level until the amounts invested were recovered and they were responsible for the operation of drinking water supply (Collins 2008).

With the Cooperation Law for Drinking Water Allocation to Municipalities (1956), the federal government agreed to finance part of the works needed without requiring fund recovery as long as the population accepted the quotas fixed by the federal authority, previous relevant socioeconomic studies. Two facts stand out in this evolutionary process of water policy of Mexico. On the one hand, the gradual and paradoxical centralization at federal level, of the functions associated with the growth promotion of agricultural production from the control of water for irrigation, and the growing interference in the provision of drinking water and sewerage services, given the implicit renunciation of local authorities to exercise their powers. On the other hand, it is observed the emergence of the fee dispersion and its lack of updating, as a phenomenon that decades later would become a powerful reason to promote the decentralization of functions in local drinking water supply and return them to municipalities.

Water institutional process continued on a path that consolidated the presence of federal authorities at local level, which is confirmed by the creation of the Ministry of Human Settlements and Public Works (SAHOP in Spanish), whose duties consisted of infrastructure construction and its operation. In 1980, it was tried to reverse the tendency towards centralization of functions in drinking water and the provision of related services, by presidential decree that returned to the states and municipalities such obligations. However, this first attempt was not successful; federal water responsibilities continued to be part of institutional functions of the SAG, the SRH and, later, the Ministry of Urban Development and Ecology (Sedue in Spanish).

It is not until 1983, with the amendment of Article 115 of the Constitution, that a process of decentralization of drinking water and sewage begins, which was institutionally strengthened with the creation of the National Water Commission (Conagua in Spanish) in 1989 and the creation of the Operating Organizations of Drinking Water, Sewage and Drainage (OOAPAS) in the early 1990s (Soares 2007).

Water policy transition in Mexico has two essential aspects to put contemporary functioning of water markets in the country into perspective. On the one hand, the fact that with the process of centralization of functions in water, a loss of local technical capacities for the implementation of a strategy for efficient water resources use was sponsored. On the other hand, a public culture was promoted, which considered that federal authority presence was essential and indispensable in the process of providing drinking water services to localities, and a context that assumed that market access had low cost, derived from the lags in updating consumption rates.

The fragility of public finances from the 1980s, increasing costs of the operation model of water policy, and a better understanding of the ecosystem implications of water use, provide the scenario in which the decentralization of operations and economic, fiscal, and financial instruments are the axes that articulate the contemporary water strategy.

By considering water markets as spaces of interaction based on water volume transactions, a way to set the different areas covered by that name is the one originated from the classification of the Public Registry of Water Duties (REPDA in Spanish). This institutional arrangement is essential in the design of water management mechanisms, as its update, from the 1990s, allows to incorporate at the same time different dimensions of the water management issues, not only in terms of demanded volumes, but also the resilience and pressure on water resources and, of course, the financial aspects.

In the process of water management evolution, which involves the issuance of grants, allocations and permits from the federal authority to attend multiple purposes such as drinking water supply (through OOAPAS), productive use in food production, industrial activities, power generation, and other uses, such as recreational and environmental, it gradually accumulated an information lag on rights and utilization volumes. The REPDA organizes information of water exploitation permits and it sets limits to applicants; these authorizations can be assigned to individuals or entities, according to aquifer availability and the intended water use (Garduño 2003).

Water supply required for various uses can occur in various ways. In the case of productive uses and energy production, they can be accomplished by direct exploitation by production or consumption units, up to the limits set by the corresponding authorizations. In the case of massive drinking water supply in urban areas, usually the authorization is conducted through the OOAPAS or private water services companies, if available at local level, although it can be requested authorization to directly exploit federal waters. All uses and sources, including sewage discharge, have a record in the Public Registry of Water Duties.

The REPDA provides an adequate approximation to the characteristics of water transactions for different uses and it puts into perspective the financial aspects of the water management structure in the country. However, REPDA ascertains information of authorized volumes and not of the actually used, which may be greater than those authorized, subject to penalties or cancelation of the operation, or below the set limits. These differences have created a market to buy and sell rights, prior approval of water authorities.

7.2 Contemporary Water Markets in Mexico and Some of Their Main Features

Water markets are defined herein as the area of interaction of water supply and demand for generic uses considered by REPDA. Such regulated and monopolistic markets have the particularity that the supply is limited by the amount of the corresponding authorizations according to its use and sources availability. The volume of records currently valid in the REPDA exceeds 450,000 and, in volumetric terms, it implies an allocation of approximately 88 % of surface water and 12 % of groundwater in the country (Conagua 2011a).

Table 7.1 puts into perspective the existence of different water transaction areas; the General Regime item, integrated by extractive consumption done by industry, draws attention.

As it can be seen, water markets can be segmented into three groups: water for production, water for people, and water for energy; of course, a set of specific activities is associated with such categories.

The amount of duties that are assigned to the uses is determined based on the characteristics of availability segmented in nine zones, in the declared uses, and authorized levels in each concession.

It is important to distinguish between water policy objectives and achieving objectives by drinking water policy performers, especially in urban centers. However, it is necessary to order a little the problem of tariffs. Perhaps because of the way the Mexican institutional design has evolved over time, one may have the impression that the main problem of the national water policy is to achieve financial strengthening of the drinking water supply systems in human settlements, as a central factor to reduce pressure on water resources. However, due to consumption distribution and its influence on water resources, increasing the magnitude of the payments made by domestic water consumers—important and relevant matter—it does not seem to be the issue that solves by itself, the problem of supply system sustainability or pressure on water stocks.

Table 7.1 Duties for the use of national waters, according to availability zone, 2010 (cents per cubic meter)

Use	Zone								
	1	2	3	4	5	6	7	8	9
General regime	1828.94	1463.1	1219.24	1005.89	792.48	716.23	539.09	191.53	143.54
Drinking water, consumption more than 300 l/inhabitant/day	72.46	72.46	72.46	72.46	72.46	72.46	33.74	16.85	8.39
Drinking water, consumption equal to or less than 300 l/inhabitant/day	36.23	36.23	36.23	36.23	36.23	36.23	16.87	8.43	4.19
Agricultural, without exceeding the assigned volume	0	0	0	0	0	0	0	0	0
Agricultural, for every m^3 that exceeds the assigned volume	12.95	12.95	12.95	12.95	12.95	12.95	12.95	12.95	12.95
Spas and recreational centers	1.04	1.04	1.04	1.04	1.04	1.04	0.51	0.24	0.11
Hydropower generation	0.38	0.38	0.38	0.38	0.38	0.38	0.38	0.38	0.38
Aquaculture	0.3	0.3	0.3	0.3	0.3	0.3	0.15	0.07	0.03

Source Conagua, Statistics on water in Mexico, 2011 edition

National water strategy, in an environmental context as proposed by the General Law of Ecological Equilibrium and Environmental Protection (LGEEPA in Spanish), has as main objectives to ensure citizen welfare, to encourage economic prosperity and to promote its preservation (SEMARNAT 2008). To encourage water allocation processes in such large set of goals is not a matter of relative prices. Economic instruments available for its implementation are of three types: fiscal, financial, and market. The first type concentrates on the amounts and collection of rights. The second type, depending on water use, in the securities system and liability insurance and third type, in the operation of a market of rights transfer.

However, from the aforementioned instruments, the one that affects the operational capacity of water policy is the payment of rights. In this regard, it stands out that the primary sector, which has the major direct water consumption in the country, generates an inversely proportional contribution on rights payments (Table 7.2). This, although subject to controversy from a water perspective, is not different from the experience in other countries. When considering a transversality approach, as is done in the water footprint analysis and it is explained throughout this study, we realize that water content used directly in the manufacture of products does not necessarily reflect the importance that this good has as production input. The use of indicators of direct use prevents to quantify the water volumes that are actually utilized in the manufacture of consumption goods and the possible existence of an underlying cross-subsidies mechanism (Table 7.3).

In a scenario that only considers the direct consumption of water users, it might seem that the essential problem is the use efficiency in the primary sector. This is appropriate if sectoral linkages are not considered, but also heterogeneity in the productive capacities of the federal states (Fig. 7.1).

It is possible that because drinking water supply and the intrinsic value that it may have for the consolidation of an initiative of decentralization of functions in this area are important for local governments, there is a greater coverage about budget shortfalls faced by national water sector in relation to drinking water, sewerage and sanitation subsector (APALS in Spanish). And while it is certainly an important component of federal budgetary disbursements, it does not seem to correspond to, in volume terms, the magnitude of national consumption for domestic purposes, with the importance acquired by the financing derived from the collection of domestic rates.

The growing and repeated emphasis on national dispersion of tariffs for domestic uses, the relative opacity in the criteria for establishing levels in local areas, and heterogeneous coefficients of collection efficiency at local level, are matters of water agenda that have contributed to create a relative confusion in public opinion.

It is true that OOAPAS must maintain a financial position to strengthen its capacity to modernize its infrastructure, expand coverage, maintain and ensure supply quality, this must be done based on consumption payments. But one cannot forget that this type of consumption is only part of the volume of total domestic waters and also, once the flows to be used by local supply systems are authorized, there are several uses: residential, commercial, and industrial.

Table 7.2 Collection of CONAGUA of duties by concept, 2000–2009 (millions of pesos at constant 2009 prices)

Concept	2000	2001	2002	2003	2004	2005	2006	2007	2008	2009
Use of national waters	7343.80	7122.30	7645.80	8231.50	7796.50	7814.40	7386.80	7875.10	8003.70	7938.50
Bulk water supply to urban and industrial centers	1311.70	1332.90	1292.80	1473.90	1383.90	1634.60	1516.40	1601.50	2148.50	2074.70
Irrigation services	168	192.5	193.3	176.2	179.6	184.3	176.4	210.2	204.8	225.7
Material extraction	46.4	50.1	38.8	34.9	44.3	40.7	60.1	40.1	44.9	45.7
Use of receiving bodies	51.1	91.2	71	82.1	80.8	61.4	55.7	63.4	61.2	179.4
Use of federal zones	29.3	28.4	28.3	30.2	38.6	32.5	30.6	38	33	38.2
Various	330.4	275.6	267.6	133	89.8	89.9	134.1	103.8	348.6	213.9
Total	9280.70	9093.00	9537.60	10161.80	9613.50	9857.80	9360.10	9932.10	10844.70	10716.10

Source Conagua, Statistics on water in Mexico (2011c:87)

Table 7.3 Volumes and surfaces allocated by type of use and exploitation at December 31, 2010

Use	National waters				Water discharges		Federal zones		Material extraction	
	Titles	Extraction volume m³/year			Sewage		Titles	Surface in m²	Titles	Volume
		Surface water	Groundwater	Total	Titles	Discharge vol. m³/year				
Agricultural	145,113	36,049,022,148	17,708,607,516	53,757,629,664	325	6,899,354	60,080	1,257,544,652	4	33,208
Agroindustrial	66	464,920	5,134,357	5,599,277	18	630,360	5	153,471		
Household	15,462	13,652,848	25,809,992	39,462,840	77	2,360,219	1,404	697,898		
Aquaculture	876	1,035,474,699	18,083,070	1,053,557,769	370	3,898,126,148	234	4,969,584	1	1,095
Services	5,616	492,361,927	746,169,024	1,238,530,951	6,246	942,141,089	6,221	43,255,878	1,379	72,483,865
Industrial	4,897	4,322,463,210	1,428,620,946	5,751,084,156	2,295	10,184,638,775	492	10,330,153	108	7,670,910
Livestock	35,780	63,536,621	127,310,243	190,846,864	1,113	18,582,363	9,787	534,710,056		
Urban public	120,851	4,266,958,691	7,084,807,514	11,351,766,205	3,120	3,395,041,432	18,015	3,489,784	2	2,000
Multiple	38,065	2,246,543,448	2,333,099,495	4,579,642,943	870	4,484,404,359	4,234	312,054,232	245	36,181,694
Power generation	103	165,066,111,646	778,857	165,066,890,503	0	0	23	539,465		
Trade	2	0	80,000	80,000	0	0	102	28,882	1,381	43,153,346
Others	5	300 000	241,416	541,416	0	0	270	775,228	34	80,953
Total	366,836	213,556,890,159	29,478,742,430	243,035,632,589	14,434	22,932,824,099	100,867	2,168,549,283	3,154	159,607,071

Source Conagua, Statistical compendium of water management (CEAA in Spanish) (2011a:41)

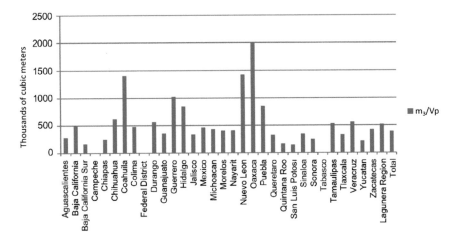

Fig. 7.1 Water volume per million of agricultural production value (Cycle 2008–09). *Source* ConaguaA (2009), Agricultural statistics of irrigation districts, Mexico

It should be emphasized that the sustainability of urban water systems is important, but it is related to the local supply security and not with the way water uses are socially assigned at federal level. The financial health of OOAPAS depends not only on the amount and magnitude of their income, but also on their cost structure. These are determined, greatly, by the technological choices of the way at federal level, water use is established and encouraged. And it compromises water security.

Sectoral investments in water supply in urban areas tend to increase (Fig. 7.2). In part, this is due to the delays accumulated in this area over time, but they are also associated with increased costs of water provision in a context of overexploitation of watersheds and an operation model with increased energy costs (Fig. 7.3).

Drinking water financing rests predominantly on the federal level, but the co-responsibility of the state and municipal governments related to contribution is tending to increase (Fig. 7.4).

Meanwhile, although a significant difference remains between billing and collection of water, sewerage and sanitation services at national level (Fig. 7.5), it does not explain the eventual financial fragility of the national water sector as a whole, but the one related to urban water supply.

Sustainability of urban drinking water supply services is not a minor issue; it is necessary that users have a reference of direct costs, but also of the intangible costs such as the supply difficulty costs and those related to ecosystemic aspects. Bills should include not only aspects of production or distribution but also those related to treatment.

For a signal strategy via relative prices to be effective in moderating consumption patterns and incorporating innovative practices to increase utilization efficiency, it must rest on transparency and agent knowledge about potential hidden

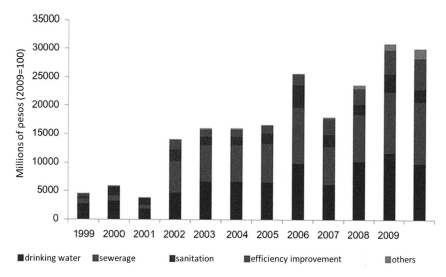

Fig. 7.2 Subsector total investments: including programs of Conagua, SEDESOL and Banobras. *Source* Data from Conagua/SGAPDS/Management of Studies and Projects of Drinking Water and Sewerage Network

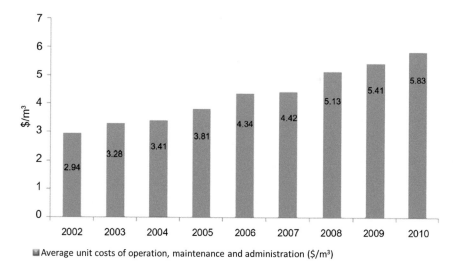

Fig. 7.3 Average unit costs of operation, maintenance, and administration OOAPAS *Source* Data from IMTA (2010), Program of Management Indicators of Operating Organizations of the Mexican Institute of Water Technology, SEMARNAT

water use subsidies. There is not enough evidence to infer that any volumetric savings that could be achieved via household consumption would compensate for the lack of incentives for water conservation in the productive sector performance.

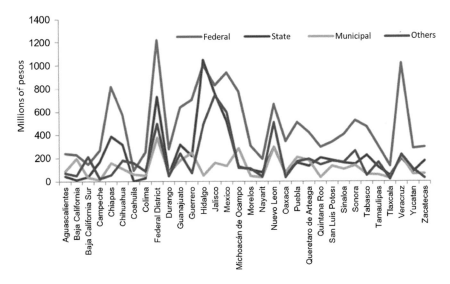

Fig. 7.4 Investments reported by federal states by resource origin sector 2011. *Source* Data from Conagua (2012), Situation of Subsector of Drinking Water, Sewerage and Sanitation

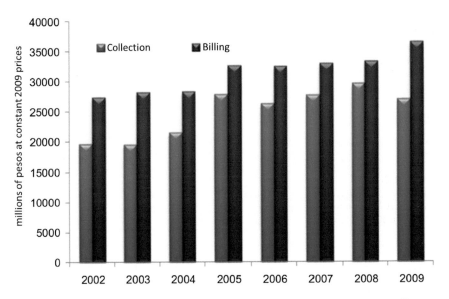

Fig. 7.5 Annual billing and collection of operating organizations. *Source* data from Conagua (2011b). Instruments for water management in Mexico

At the same time that investments in drinking water supply for public consumption has increased, it has also increased the amount of irrigation infrastructure in the country (Fig. 7.6).

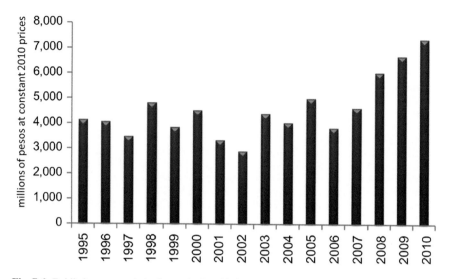

Fig. 7.6 Public investment in hydro-agricultural infrastructure. *Source* data from Statistical Annex of the fifth government report (2011)

The above statement should not be misunderstood; it is necessary for the economy as a whole to reduce the pressure on water resources and to promote use practices to be sustainable, from water, ecosystemic, and financial perspectives. Of course, it is desirable that it happens in all sectors.

References

Aboites, L. and Valeria Estrada. *Del agua municipal al agua nacional. Materiales para una historia de los municipios de México (1901–1945)*. México: CONAGUA/AHA/CIESAS/COLMEX, 2004.
Cantú, Mario and Héctor Garduño. *Administración de derechos del agua: de regularización a eje de la gestión de los recursos hídricos*, México: FAO, 2003.
Collado, Jaime. "Entorno de los servicios de agua potable en México". In *El agua potable en México. Historia reciente, actores, procesos y propuestas*, coordinated by Roberto Olivares and Ricardo Sandoval, 3–28, México: ANEAS, 2008.
CONAGUA. *Reporte económico de administración del agua*, México: CONAGUA, 2008.
CONAGUA. *Estadísticas agrícolas de los distritos de riego*, México: CONAGUA, 2009.
CONAGUA. *Reporte económico de administración del agua*, México: CONAGUA, 2010.
CONAGUA. *Compendio estadístico de administración del agua*, México: CONAGUA, 2011a.
CONAGUA. *Gaceta de administración del agua,* México: CONAGUA, 2011b.
CONAGUA. *Estadísticas del agua en México: instrumentos de gestión del agua en México*, México: CONAGUA, 2011c.
CONAGUA. *Situación del subsector agua potable, alcantarillado y saneamiento*, México: CONAGUA, 2012.
Consejo Consultivo del Agua. *Gestión del agua en las ciudades de México*, México: Consejo Consultivo del Agua, 2011.

Dau, Enrique. "Retrospectiva, análisis y propuestas para impulsar una etapa definitoria del sector agua potable y saneamiento mexicano". In *El agua potable en México. Historia reciente, actores, procesos y propuestas*, coordinated by Roberto Olivares and Ricardo Sandoval, 97–104, México: ANEAS, 2008.

Garduño, Hector. Administración de derechos de agua: experiencias, asuntos relevantes y lineamientos, FAO Legislative Study 81, 2003.

Hernández, Julia and Jesús Borrego. La economía del agua: una aproximación para su gestión responsable, México: Universidad Autónoma de Chihuahua, 2009.

IMTA. *Programa de Indicadores de gestión de organismos operadores del Instituto Mexicano de Tecnología del Agua*, México: SEMARNAT, 2010.

Pineda, Nicolás and Alejandro Salazar. "De las Juntas Federales a las empresas de agua: la evolución institucional de los servicios urbanos de agua en México 1948–2008". In *El agua potable en México. Historia reciente, actores, procesos y propuestas*, coordinated by Roberto Olivares and Ricardo Sandoval, 57–76, México: ANEAS, 2008.

Presidencia de Republica. *Quinto Informe de Gobierno, Anexo Estadístico*, México: Presidencia de la República, 2011.

SEMARNAT. *Plan Nacional Hídrico (2007-2012)*, México: SEMARNAT, 2008.

SEMARNAT. *Distribución de los costos del cambio climático entre los sectores de la economía mexicana: un enfoque de insumo-producto*, México: SEMARNAT, 2009.

Soares, Denise. "Crónica de un fracaso anunciado: la descentralización en la gestión del agua potable en México", *Agricultura, Sociedad y Desarrollo*, vol. 4, no. 1, (2007): 19–37.

Van der Zaag, Pieter and Hubert Savenije. "Water as an economic good: The value of pricing and the failure of markets." In *Value of Water Research Report* 19, The Netherlands: UNESCO/IHE/Institute for Water Education, 2006.

Vargas, Sergio, et al. *La gestión de los recursos hídricos: realidades y perspectivas*, México: IMTA, 2009.

Part II
Pressures on Water Availability, Its Use and Welfare Effects

Chapter 8
Water Use Pattern

Eugenio Gómez-Reyes

Abstract Water availability and sufficiency to adequately meet the different objectives of its use are not only related to the existence of accessible sources and utilization capacity; use practices and competition between different sectors and regions are also determinants to the apparent scarcity. A water management institutional strengthening process requires recognition of the exploitation forms that have been consolidated as a result of federal institutional evolution. References to the federal water use pattern consider blue (color) water, but the potential and incentive structures to modify the use of green and gray water are rejected.

Keywords Water uses · Consumtion patterns

8.1 Water Use

National water volume that has been allocated or assigned is registered in the Public Registry of Water Duties (Repda) (*supra* Chap. 5), classifying the uses of water in offstream (agriculture, public supply, self-supplying industry, and thermoelectric) and instream (hydropower). For instream use, the same water is turbined and counted as part of the offstream use to which it is destined. The volume of water used for hydropower does not exert stress on water resources.

According to Repda, offstream uses (UCA in Spanish) are classified into 12 groups that the National Water Commission, Conagua (2011), has grouped into the following four headings:

- Agriculture: livestock, aquaculture, multiple, and other uses; this use represents food security.
- Public supply: urban domestic and public use.
- Self-supplying industry: agro-industrial uses, services and trade.
- Electricity generation: for non-hydro power plants.

From the total volume allocated for offstream uses (Table 8.1), agriculture has the greatest volume (76.7 %), followed by public supply (14.1 %), electricity generation (5.1 %), and self-supplying industry (4.1 %). It should be noted that 63 % of the

R.H. Pérez-Espejo et al. (eds.), *Water, Food and Welfare*,
SpringerBriefs in Environment, Security, Development and Peace 23,
DOI 10.1007/978-3-319-28824-6_8

Table 8.1 Allocation volume (m³/s) for offstream uses in Mexico

HAR	Name	Agricultural			Public supply			Self-supplying industry		
		Surface	Underground	Total	Surface	Underground	Total	Surface	Underground	Total
I	Baja California Peninsula	48.453	42.84	91.292	3.266	4.598	7.864	2.283	0.729	3.012
II	Northwest	127.347	80.067	207.414	19.28	14.46	33.739	0.127	2.79	2.917
III	Northern Pacific	277.239	30.79	308.029	9.671	10.591	20.263	1.205	0.602	1.807
IV	Balsas	165.747	34.215	199.962	13.128	18.867	31.995	3.773	3.203	6.976
V	Southern Pacific	24.607	7.357	31.963	4.091	6.532	10.623	0	0.634	0.634
VI	Rio Bravo	134.64	110.699	245.339	17.377	20.104	37.481	0.381	6.374	6.754
VII	Central Basins of the North	39.415	67.859	107.274	0.222	11.542	11.764	0.032	1.998	2.029
VIII	Lerma-Santiago-Pacific	208.65	168.411	377.061	21.404	45.852	67.256	2.442	11.701	14.143
IX	Northern Gulf	93.607	26.7	120.307	11.669	5.105	16.774	13.508	1.268	14.777
X	Central Gulf	76.294	18.043	94.337	15.094	8.498	23.592	24.607	3.234	27.841
XI	Southern Border	38.559	13.413	51.972	10.401	4.122	14.523	1.459	1.839	3.298
XII	Yucatan Peninsula	2.283	49.436	51.719	0	18.645	18.645	0	15.95	15.95
XIII	Waters of Valley of Mexico	60.756	12.145	72.901	11.067	55.746	66.813	1.395	3.742	5.137
Total	Mexican Republic	1297.60	661.974	1959.57	136.669	224.664	361.333	51.211	54.065	105.277

Source Based on data from Conagua (2011)

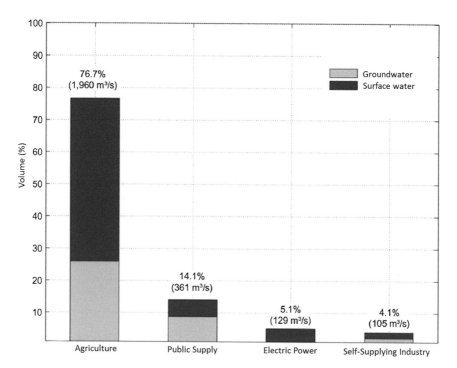

Fig. 8.1 Allocation volume for offstream uses. *Source* Based on data from Table 8.1

water used in Mexico for offstream uses comes from surface water sources (rivers, streams, and lakes); while the remaining 37 % comes from groundwater sources (aquifers), as shown in Fig. 8.1. The groundwater volume for agricultural use (662 m³/s) is greater than the sum of the total water volume allocated for the rest of the offstream uses (696 m³/s).

For public supply, most of the water comes from groundwater extraction wells, whereas about half of the water for self-supplying industry comes from underground sources. For electricity generation, water comes almost entirely from surface water.

The agricultural water use essentially refers to the water utilized to irrigate crops in 6.5 million hectares, of which 54 % corresponds to 85 Irrigation Districts and 46 % to more than 39,000 Irrigation Units for Rural Development (Urderal) (Conagua 2011). With regard to the water volume used to irrigate the Urderales, there is little official information available; there has been no thorough follow-up for these areas and much of the recorded information was lost when the institutions that supervised them disappeared (Palerm/Martinez 2009).

Public supply use includes water delivered through drinking water networks, which supply domestic users (households), as well as the different industries and services connected to these networks. Water for public supply is of the utmost importance, since social demand becomes a reflection of the right to have access to water in sufficient quantity and quality for human consumption. This water volume is directly related to the drinking water network coverage and thus to population growth, mainly, in the cities. It is estimated that 90.7 % of the population has drinking water coverage (Conagua 2011); besides, there is also distribution by pipes. It is important to note that inhabitants considered to be covered do not necessarily dispose of water of drinking quality.

Water concessions for electricity generation include the volume used in dual steam, coal-electric, combined cycle, turbo-gas, and internal combustion plants to cool and lubricate equipment and for fuel processing and turning the turbines. This volume generates 88.7 % of the electricity produced in Mexico (Sener 2010). Thermal power plants use most of the water; since they are built near surface water bodies, it turns out that most of the water allocated for electricity generation comes from surface water sources. It should be noted that 76.6 % of the water allocated to thermoelectric plants in Mexico corresponds to the coal-electric plant in Petacalco, located on the Guerrero coast, close to the mouth of the Balsas River.

In an offstream use context, the self-supplying industry is represented by industries that take their water directly from rivers, streams, lakes, or aquifers. Among the biggest consumers of water in this category are the sugar (with the highest intake), chemical, petroleum, pulp and paper, textiles, and beverages industries (Conagua 2011).

Figure 8.2 shows the way in which the volume has been allocated for offstream uses in hydrological-administrative regions (HAR). It can be observed that the HAR with more than 200 m³/s of allocation water are: VIII Lerma-Santiago-Pacific (459 m³/s), IV Balsas (339 m³/s), III Northern Pacific (330 m³/s), VI Rio Bravo (293 m³/s), and II Northwest (244 m³/s). It is worth noting that the volume allocated to these HAR represents 65 % of the total water allocated for offstream uses in the country, being VIII Lerma-Santiago-Pacific HAR the largest consumer.

It should be emphasized that agricultural use is at least 58 % of total allocations in each HAR, with the exception of the XIII Waters of the Valley of Mexico, where it represents 49 %. In the VII Central Basins of the North and III Northern Pacific HAR, it is used up to 88 and 93 %, respectively. In the VIII Lerma-Santiago-Pacific HAR, the volume of water (377 m³/s) exceeds in 38 m³/s the total volume allocated to the second consumer HAR (IV Balsas).

For public supply, the HAR that consume more water for their inhabitants are, obviously, those with the largest population: VIII Lerma-Santiago-Pacific and XIII Waters of the Valley Mexico with 67.256 and 66.813 m³/s for 21.582 and 21.141 million inhabitants, respectively.

As for electricity generation, the HAR that consumes most of the water is IV Balsas (101 m³/s), where the Petacalco coal-electric plant is located. This volume is greater than the volume that is allocated to the HAR with the highest public consumption (VIII Lerma-Santiago-Pacific; 67 m³/s) and it is almost equal to the total volume allocated to self-supplying industry (105 m³/s). It should be noted that the use of water for electricity generation is the only offstream use that does not has water concessions in all of the HAR of the country, that is, there are no concessions in the III Northern Pacific, V Southern Pacific, and XI Southern Border HAR.

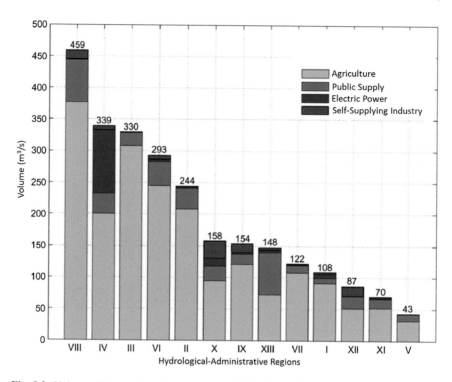

Fig. 8.2 Volumes allocated for offstream uses in HAR. *Source* Based on data from Table 8.1

For self-supplying industry, the X Central Gulf HAR has the largest water concession (27.841 m^3/s), while the V Southern Pacific has the lowest water consumption (0.634 m^3/s). Even when the offstream water use for this heading represents only 4.1 % of the total water use in the country, it is present in all the hydrological-administrative regions.

Therefore, the distribution of the volume allocated for offstream uses in the country (Fig. 8.3) indicates that agricultural use predominates, followed by public supply use. In the latter case, consumption is emphasized in the metropolitan areas of population centers with more than 500,000 inhabitants: Mexico City, Guadalajara, Monterrey, Puebla, Tijuana, Leon, Juarez City, Xalapa, Veracruz, Villahermosa, Tuxtla Gutierrez, Oaxaca, Hermosillo, Durango, Reynosa, Cancun, San Luis Potosi, Merida, Tampico, Culiacan, Cuernavaca, Acapulco, Chihuahua, Morelia.

Water distribution for electricity generation and self-supplying industry is concentrated in specific places, for example, in the Balsas Basin where the Petacalco coal-electric plant is located and in the northern, central, and southern industrial corridors.

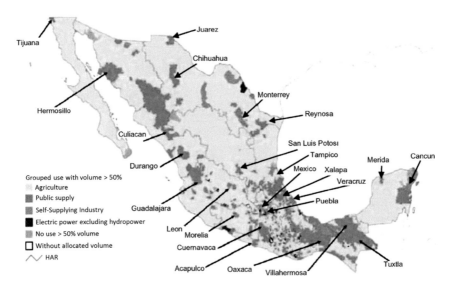

Fig. 8.3 Distribution of volume allocated for offstream uses in Mexico. *Source* Adapted from Conagua (2011)

8.2 Degree of Stress on Agricultural Use

Similar to the degree of stress exerted on water resources (percentage of water used for offstream uses compared to the renewable water resources), it can be used an index that expresses the degree of stress on agricultural water use, i.e., the percentage that represents the water allocated in other offstream uses with respect to agricultural water use. Thus, it can be considered that if this percentage is greater than 40 %, there is a strong stress on agricultural use. Additionally, it can be identified the water use that exerts the highest stress on agricultural use, since each of the water uses is a partial component of the degree of total stress on the agricultural use (Table 8.2).

Nationally, Mexico experiences a degree of stress on agricultural use of 38.7 % on average (Table 8.3), which is considered a moderate level, near to a high level. The stress (24.6 %) is mainly exerted by the public supply use. These water uses are in conflict because public supply is increasingly required by population growth in the country, while agricultural use is required for the food security of the growing population. This antagonistic stress condition can be reversed if both water uses share the same water, that is, residual water from public supply can be treated for irrigation. Then there would be no competition for water resources and the use generated by population growth and food security would be complementary.

In the case of the XIII Waters of the Valley of Mexico HAR (Fig. 8.4), there is the highest degree of stress on the agricultural use of water (103 %), caused by the public supply use (92 %). The III Northern Pacific HAR has the lowest degree of stress on agricultural water use (7 %). On the other hand, the other offstream uses exert stress

Table 8.2 Allocation volume (m³/s) for electricity generation in Mexico by source

HAR	Name	Electricity generation			Total		
		Surface	Underground	Total	Surface	Underground	Total
I	Baja California Peninsula	0	6.31	6.31	54.002	54.477	108.479
II	Northwest	0.222	0	0.222	146.975	97.317	244.292
III	Northern Pacific	0	0	0	288.115	41.984	330.099
IV	Balsas	98.998	1.522	100.52	281.646	57.807	339.453
V	Southern Pacific	0	0	0	28.697	14.523	43.22
VI	Rio Bravo	1.681	1.871	3.551	154.078	139.047	293.125
VII	Central Basins of the North	0	0.888	0.888	39.669	82.287	121.956
VIII	Lerma-Santiago-Pacific	0	0.666	0.666	232.496	226.63	459.126
IX	Northern Gulf	1.903	0.19	2.093	120.687	33.264	153.951
X	Central Gulf	11.733	0.222	11.955	127.727	29.997	157.725
XI	Southern Border	0	0	0	50.419	19.375	69.793
XII	Yucatan Peninsula	0	0.285	0.285	2.283	84.316	86.599
XIII	Waters of Valley of Mexico	0.698	2.156	2.854	73.916	73.789	147.704
Total	Mexican Republic	115.233	14.111	129.344	1600.71	954.814	2555.52

Source Based on data from Conagua (2011)

Table 8.3 Degree of stress on agricultural use in Mexico (percentages)

HAR	Name	PS[a]	SSI[b]	EG[c]	Total	Classification
I	Baja California Peninsula	8.6	3.3	6.9	18.8	Low
II	Northwest	16.3	1.4	0.1	17.8	Low
III	Northern Pacific	6.6	0.6	0	7.2	No stress
IV	Balsas	16	3.5	50.3	69.8	High
V	Southern Pacific	33.2	2	0	35.2	Moderate
VI	Rio Bravo	15.3	2.8	1.4	19.5	Low
VII	Central Basins of the North	11	1.9	0.8	13.7	Low
VIII	Lerma-Santiago-Pacific	17.8	3.8	0.2	21.8	Moderate
IX	Northern Gulf	13.9	12.3	1.7	28	Moderate
X	Central Gulf	25	29.5	12.7	67.2	High
XI	Southern Border	27.9	6.3	0	34.3	Moderate
XII	Yucatan Peninsula	36.1	30.8	0.6	67.4	High
XIII	Waters of Valley of Mexico	91.6	7	3.9	102.6	Very high
Average	Mexican Republic	24.6	8.1	6	38.7	Moderate

Source Based on data from Table 8.1
[a]*PS* Public Supply
[b]*SSI* Self-Supplying Industry
[c]*EG* Electricity Generation

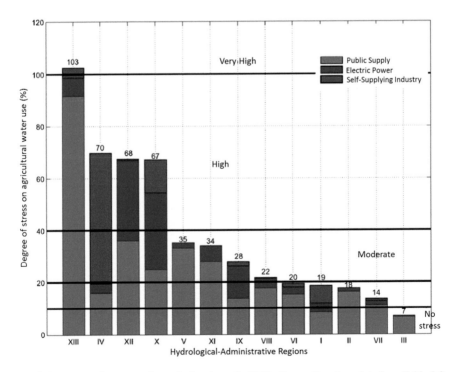

Fig. 8.4 Degree of stress on the agricultural use for HAR. *Source* Based on data from Table 8.2

differentially on the agricultural use. In the IV Balsas HAR, the stress is exerted by the use of water for the Petacalco coal-electric plant (self-supplying industry); in the XII Yucatan Peninsula HAR, public supply and electricity generation exert the stress; while in the X Central Gulf HAR, the stress comes from the three offstream uses: public supply, electricity generation, and self-supplying industry.

8.3 Conclusions

In order to cope with the stress of agricultural water use in the coming years, it will be necessary to take actions to avoid competition with the public supply use. A very important aspect to consider in future scenarios of Mexico is the increase in population and its concentration in urban areas, which increases the degree of stress on the agricultural use. The increment in water treatment and wastewater reuse is a favorable alternative to reduce this stress. On the other hand, to ensure economic development and to reduce the stress on the agricultural water use, the industry that renovates the water used in their production processes (for example, hydropower, geothermal, solar, and wind power) should be favored.

References

* indicates internet link (URL) has not been working any longer on 8 February 2016.

CONAGUA. *Estadísticas del agua en México.* México: SEMARNAT, 2011.
Palerm, Jacinta and Tomas Martinez. *Aventuras con el Agua. La administración del agua de riego: historia y teoría,* México: Colegio de Posgraduados, 2009.
*SENER. *Cuarto Informe de Labores 2010* http://www.sener.gob.mx/webSener/res/0/Cuarto InformeLaboresSENER2010.pdf, Accessed October 15, 2010.

Chapter 9
Change in the Dietary Pattern and Water Security

Andrea Santos-Baca

Abstract Food production and consumption are key activities in our societies. Paradoxically, despite the technological progress over the past 100 years, it has not been able to secure sufficient and healthy feeding for everyone. The validity of food crisis has placed the different dimensions of food consumption at the center of attention. The aim of this chapter is to relate theoretically and empirically, three of the considered dimensions: food sovereignty, water security, and change in the pattern of food consumption. In the first part, the theoretical discussion of the relationship of these dimensions is presented, placing at the center of the argument the concept of food security. The second part contains an exercise that estimates the effect of changes in food consumption, food sovereignty, and water security in Mexico in 1992 and 2010, using data from the national survey of household income and expenditure. The results indicate the importance of valuing more than one of the dimensions involved in the realization of the food right.

Keywords Food consumption patterns · Food sovereignty · Water security · Free trade

9.1 Food Sovereignty, Water Security, and Changes in the Dietary Pattern

In this section, the relationships between food security and sovereignty and their relation to changes in the dietary pattern and water use are presented.

9.1.1 The Debate on Food Security

Food security is a multidimensional concept that goes back to the 1970s in the context of the threat of a global food crisis. The initial focus in the World Food

© The Author(s) 2016
R.H. Pérez-Espejo et al. (eds.), *Water, Food and Welfare*,
SpringerBriefs in Environment, Security, Development and Peace 23,
DOI 10.1007/978-3-319-28824-6_9

Conference in 1974[1] (FAO 2003: 45) centered on the availability problems and, with minor modifications, at the 1996 World Food Summit, the most complete definition of food security was created:

> Food security exists when all people, at all times, have physical and economic access to sufficient, safe and nutritious food that meets their dietary needs and food preferences for an active and healthy life (WFS 1996).

The international community gathered in Rome negotiated a series of agreements on the multiple factors that influence the right for food; these agreements are characterized by their complexity, ambiguity, and ambition.[2] However, the effective response focused on a single goal: to reduce by half the world's population in a poverty situation by 2015 (FAO 2003: 47).

In 2009, the World Summit on Food Security (WSFS) was held in the context of the main food international prices rising; as a result, the declaration of 2009 calls to halt the increase in the number of people suffering from hunger, malnutrition, and food insecurity. Inadequate efforts to achieve the goals of the 1996 Summit are recognized.

The 1996 Summit expanded the food security concept, but considered free trade as the most efficient way to achieve it; it is a significant difference from the food security concept in 1974, which noted: "[…] All States should strive to the utmost to readjust, where appropriate, their agricultural policies to give priority to food production, recognizing […] the interrelationship between the world food problem and international trade." (WFC 1974: 11). By contrast, the 1996, 2002, and 2009 declarations establish that trade is a key element to achieve global food security, a premise that intercalates in the 2009 Summit with the reappreciation of local food production (WSFS 2009: 5).

Facing a food, health, and environmental crisis and a complete failure of the 1996 Rome Declaration, the Via Campesina social movement demands food sovereignty. The notion that opposes the wrong growing use of the food security concept that has favored the power of the largest agro-food companies, promotes free market and does not consider where the food comes from, by whom, and how are produced (GRAIN 2006: 2). It is declared that trade liberalization has imposed an agriculture where farmers have no place, because transnational companies

---

[1]The Universal Declaration on the Eradication of Hunger and Malnutrition of 1974 states that "The well-being of the peoples of the world largely depends on the adequate production and distribution of food […] establishment of a world food security system which would ensure adequate availability of, and reasonable prices for, food at all times, irrespective of periodic fluctuations […] of weather […] and should thus facilitate […] the development process of developing countries" (CMA 1974).

[2]Some of the commitments were to: eradicate poverty, strengthen peace, ensure gender equality, promote national and international solidarity, restrain 'excessive' rural–urban migration, have a proper diet, have food safety, have access to all health services and education, secure and lucrative jobs, equitable access to productive resources, sustainable development, recovery of traditional knowledge of indigenous communities, rural development, technological transfer and development, reducing damage from natural disasters, etc. (WFS 1996).

control almost all of the food chain (GRAIN 2006: 2), control that is the main cause of failure of the 1996 Summit commitments.

Food sovereignty has two meanings: it is the right of people to healthy and culturally appropriate food produced through ecologically sound and sustainable methods and their right to define their own food and agriculture systems (Nyéléni 2007: 1) and to define their own agricultural and food policies (FFS 2002: 1). It does not oppose international trade, but defends the option of formulating sovereignly trade policies and practices that serve better the rights of the population (FFS 2002: 14).

9.1.2 Food Sovereignty and Water Security

Agri-food is the biggest economic sector of the world, involving more transactions and employing more people by far than any other (GRAIN 2011). This places it at the center of ecosystem deterioration, not only because it is the most affected by environmental crisis, but because it is one of the main drivers behind it (GRAIN 2009).

The type of food consumed, the manner and place in which they are produced, and the process by which they reach the tables of consumers are factors that determine the use, wear, and contamination of natural resources. It is estimated that the global agri-food system contributes about half of all human-generated greenhouse gas emissions (GRAIN 2011: 2).

In the definition of food security, concern for the environment and food is found in the supply and temporality dimensions, i.e., it is required a sustainable access and use of resources to ensure a sufficient food supply at all times. But in recent years, it has been recognized the need to add, explicitly, an environmental sustainability dimension to the definition of food security (Cuéllar 2011: 5).

The attention to water resources, crucial in food production, has increased in recent years. About 70 % of all freshwater is allocated to the agricultural sector (Ercin et al. 2011: 722), and about 90 % of the water requirements of an individual is related to food production (Liu/Savenije 2008: 888). The right to food requires water availability in sufficient quantity and adequate quality, and to increase its productivity in the agri-food sector (FAO 2012: 1).

Studies of virtual water and water footprint (WFP) recommend that regions with low water availability can mitigate this shortage by importing water-intensive food (Liu/Savenije 2008: 896). The Food and Agriculture Organization (FAO) suggests that countries whose food consumption depends on local agriculture have an increased risk to suffer food insecurity due to lack of water (FAO 2012: 2).

In the food sovereignty definition, the relationship between environment and food is presented explicitly and it proposes to modify agri-food production toward a different production model, agroecology, which means to abandon intensive and large-scale production, use of agrochemicals, monoculture, and genetically modified organisms. In relation to water, it raises the need for "a truly democratic approach for the water resources management," "to return to the traditional sustainable means to access and manage water," and "to ensure a more efficient and sustainable use of resources water" (FFS 2002: 7–8).

9.1.3   Changes in Dietary Patterns, Food Sovereignty, and Water Security

A dietary pattern is the set of products that a society considers appropriate to meet their dietary needs at a given historical moment. It is characterized by a particular combination and proportion of different sources of energy (calories) that do not correspond to a nutritional rationality or cost–benefit nor a social arbitrariness.

Changes in a dietary pattern may be: (1) substitution between food groups: legumes for animal products; (2) replacement within a food category: milk for egg within animal products; (3) substitution within a product category: substitution of pork for beef; and (4) replacement of agricultural products for industrial products (Malassis/Ghersi 1992: 66).

Changes in a dietary pattern can be explained by eight variables: (1) budget constraint: income and prices; (2) information availability; (3) time availability; (4) working and life conditions: duration and intensity of working day, urbanization degree, and transportation time; (5) goods production and trade: food market, agricultural public policy, free trade, etc.; (6) social differentiation pressure; (7) cultural constraints; and (8) need restriction. This list of variables is neither exhaustive nor presented in a hierarchical order.

Food preferences or consumption habits are determined socially and culturally and they enter in food security and sovereignty with two attributes: sustainability and nutrition. In recent years, it has been recognized the significant impact of food consumption patterns on the water needs of a society (Liu/Savenije 2008: 887). The amount of water used for food depends, to a large extent, on the dominant dietary pattern, and its changes will affect the pressure exerted on water and other resources needed in food production.

The food transition that occured in the twentieth century, first in the developed countries and then in the periphery, has had a strong impact on environment. This transition, known as "postwar food regime" (Friedmann 1994) and led by the United States, is extremely intensive in natural resources and is characterized by the replacement of cereals and legumes for animal products as the main source of energy and protein (Cepede/Lengelle 1953).

The current global context of the new food system has driven changes in the dietary patterns of societies that affect food sovereignty because of the pressure it exerts on natural resources and nutrition quality.

The World Health Organization (WHO) refers to the health changes in the developing countries and poor populations as the "double burden of diseases," a combination of hunger and malnutrition with a rapid onset of chronic-degenerative diseases that have become a serious epidemic (WHO-FAO 2003: 20). The qualitative deterioration of food intensifies the contradiction among populations "forced to underconsumption" and "forced to overconsumption." But now, underconsumption does not present itself only as "absolute hunger," but also as "hunger relative" and "negative hunger," in the form of cheap food and cheap or empty calories (Araghi 2009).

9.2 Virtual Water, Food Sovereignty, and Change in the Dietary Pattern in Mexico

In recent years, Mexico has faced different problems related to food: degradation, overexploitation and pollution of water resources, obesity and diabetes, as well as an increase dependency on food imports, among others.

Based on a household survey, the effects of dietary pattern changes on food sovereignty and water security are discussed to estimate the impact of the dietary pattern modification and increment of food imports in water requirements per capita and at national level.

9.2.1 Characteristics of Food Consumption Changes in Mexico

Food consumption characteristics can be analyzed using two approaches: apparent demand and surveys of household income and expenditure. This research chooses the latter because it allows to build dietary patterns from household consumption perspective. The years 1992 and 2010 are analyzed using information from the National Survey of Household Income and Expenditure (ENIGH in Spanish) made by the National Institute of Statistics and Geography (INEGI in Spanish), which is representative at the national level.[3]

Food was divided into nine groups, from which consumption was reduced in five of them: legumes (−25 %), dairy (−13 %), cereals (−8 %), fruits (−7 %), and tubers (−4 %); and increased in four of them: beverages (204 %), vegetables (21 %), egg (16 %), and meat (15 %) (ENIGH 1992–2010). The trend of postwar food pattern— the increase in animal products over cereals and legumes—is still present in the Mexican dietary pattern. This trend 'replaces' cheap agricultural calories (cereals and legumes) for expensive agricultural calories (animal products, fruits, and

---

[3]The following changes were made in order to handle a single unit of measure: milk and fermented milk drinks: 1 l = 1.032 kg; bottled beverages: 1 l = 1 kg. Differences were validated using mean difference tests (test T) with a confidence level of 95 %. The quantities obtained are the ones consumed per capita per year for a possibly nonexistent "average Mexican home."

vegetables); it is striking the fall in consumption of dairy and fruits, as well as the increase in beverage consumption.

Beans are the main legumes consumed in Mexico; historically, it constituted the complement of corn, so that the maize–bean combination provided most of the energy, protein, and fiber in the Mexican diet. In the period of study, bean consumption presents a significant reduction.

The intake reduction of milk explains the decrease in the dairy group consumption, although cheese, cream, and fermented milk beverages increased significantly. In Latin America, the proportion of energy obtained from cereals declined from 52 % in 1995 to 45 % in 1999 (Bermúdez/Tucker 2003). The fall in cereal consumption is explained by the reduction in corn consumption and, particularly in corn products other than tortillas; despite this, maize is the main cereal consumed in the country. Other studies show that wheat consumption—especially in the form of white flour—has increased in relation to other cereals (Ocampo/Flores 1992), while the consumption of tortillas, rice, bread, and pasta remained constant.

The survey contains information about 18 different fruits; the top three are: banana, apple, and orange, which represented 60 and 40 % of total fruit consumption in 1992 and 2010, respectively (ENIGH 1992–2010). Tubers are the last food group that showed a reduction; potato, the main food of this group, presented a drop in consumption.

Beverages are the 'food' group whose consumption increased more; this is due to the increase in nonalcoholic beverages, especially bottled water and soft drinks. Soft drinks and other processed beverages are called "empty food" because their only function is to supply energy in the form of carbohydrates, without any other nutrient. The rise in soft drinks consumption reflects the impoverishment process of the diet because the refined sugars and flours are cheap sources of energy and they are assimilated more rapidly by the body (Ocampo/Flores 1992: 270). It is possible that the reduction in milk consumption is associated with the increased consumption of soft drinks (Rivera et al. 2008: 175).

Vegetable consumption showed a significant increase; however, consumption of vegetables and fruits only represented 50 % of the recommended amounts for this type of food that provides vitamins.[4]

Meat consumption has been considered as an element representing development and modernization of societies, as well as an expression of their well-being; however, excessive consumption causes health problems associated with chronic-degenerative diseases (cancer, diabetes, and heart problems) and environmental degradation. The study of the WHO-FAO (2003) notes that "the number of people fed in a year per hectare ranges from 22 for potatoes and 19 for rice to 1 and 2,

---

[4]Fruit and vegetable consumption recommended is 400 g per capita per day, a level that is associated with a lower incidence of obesity, cardiovascular disease, diabetes type II, and different cancers (Ramírez-Silva et al. 2009: 575). The WHO indicates that at present, only a minority of the world population consumes the recommended amounts of fruits and vegetables (WHO-FAO 2003: 34).

respectively, for beef and lamb" (WHO-FAO 2003: 32). In Mexico, there has been an increase in the consumption of processed meats (sausages) and poultry, which is the main meat consumed in Mexico since the 1980s; the consumption of beef and pork has reduced.

The change in food consumption of Mexican households from 1992 to 2010 is not a dietary transition in itself, but a substitution between food groups: there is an increase in beverages, sausages, meats, poultry, egg, and vegetables; there is a decrease in cereals, legumes, milk, and fruits. There is also substitution within food categories by reducing milk consumption in dairy products and corn consumption in cereals, and increasing processed meat consumption in meat.

Modification of dietary pattern is associated with the increment in chronic-degenerative diseases that the country is facing; from 1999 to 2006, Mexico had the highest growth, worldwide, in overweight and obesity, and deaths caused by diabetes increased to 843,654 in the period 2000–2012 (EPC 2012). While there are multiple factors causing obesity and diabetes, various studies identify an obesogenic environment after the trade opening, which is closely related to the dietary changes presented: increased consumption of soft drinks, processed meats, dairy products, and refined flour (EPC 2012; Clark et al. 2012; Santos-Baca 2012).

9.2.2 Virtual Water of the Mexican Dietary Pattern

Virtual water (VW) is defined as the water volume used to produce a unit of food in the place where it was produced or, alternatively, the water volume that would be required to produce food at the site where it is consumed (Hoekstra/Chapagain 2007: 36) (Table 9.1).

Table 9.1 Summary of different scenarios

Mexico	1992		2010	
	Per capita (m^3/year)	National (million m^3/year)	Per capita (m^3/year)	National (million m^3/year)
Scenario 1	698	60,647	678	76,121
Hypothetical scenario A			698	78,399
Scenario 2	660	57,323	641	72,080
Hypothetical scenario B			677	76,012

Source The authors
Scenario 1: Food sovereignty 1992 and 2010
Hypothetical scenario A: Food sovereignty without change in dietary pattern 2010
Scenario 2: Trade openness 1992 and 2010
Hypothetical Scenario B: Trade openness without NAFTA

The estimate of the water amount in the main food consumed and produced in Mexico was based on the calculations of WFP (blue, green, and gray) of Mekonnen/Hoekstra (2010a, b), making equivalences between the classifications used in Mexico and those used by them (Table 9.2). Due to these equivalences, the results obtained may be under- or overestimated, particularly processed food with high VW (Hoekstra/Chapagain 2007: 39) may be underestimated. Data from Ercin et al. (2011) was used for soft drinks, modifying the source of the sweetener, and utilizing data for Mexico.

VW analysis of the main food consumed in Mexico is analyzed from two scenarios; in the first one it is assumed that all food is produced internally, i.e., there is food sovereignty (Scenario 1). This leaves out the effects of international trade on the water amount that is actually used in the food consumed according to the country of origin. The second scenario considers the international trade effects with the assumption that all food imports come from the United States. In both scenarios, the food VW is the same in both years; it is the average of the period 1996–2005, ignoring changes in productivity and technology in food production. The increase in population, which is very important to understand the magnitude of VW associated with the food diet, is considered in both scenarios. Additionally, two hypothetical situations corresponding to 2010 are presented in each scenario: food sovereignty without change in the dietary pattern and free trade, but without the North American Free Trade Agreement (NAFTA). The objective is to estimate the impact that dietary pattern changes and increasing dependence on food imports have on water requirements per capita and at national level.

9.2.2.1   Scenario I: Food Sovereignty

In Scenario I, it is assumed that all food consumed in Mexico is produced internally. In the analysis period, the per capita water requirement for food (CWRF) declined 2 %, from 697.9 m^3 in 1992 to 677.6 m^3 in 2010.[5] This reduction is due to changes in dietary patterns, especially to the decrease in consumption of beef and milk, both being water-intensive products, and an increased consumption of processed products (soft drinks, sausages, dairy drinks), whose VW is less or may be underestimated because of lack of information.

In 2010, five types of food accounted for 74 % of the total CWRF: tortilla (20 %), beef (20 %), milk (18 %), poultry (8 %), and egg (7 %). The most water-intensive foods are beef (14 m^3 per kg), cheeses (8.2 m^3 per kg), ham (8 m^3 per kg), pork (7.4 m^3 per kg), and sausages (6.8 m^3 per kg). However, by volumes consumed, only beef contributes significantly to the total virtual water.

---

[5]According to Liu/Savenije (2008), the per capita water requirement for food in China was 255 m^3 in 1961 and 860 m^3 in 2003. Hoekstra/Chapagain (2007) notes that in the United States it is 806 m^3. Both numbers come from an apparent demand approach.

Table 9.2 Virtual water content of the main food consumed in Mexico and produced locally

Food	Virtual water Mexico 1996–2005 (m³/ton)		Green water	Blue water	Gray water	Total
	Equivalent HS Code					
Tortilla	1005 y 110220	Corn and corn flour	1,876	63	362	2,300
Bread	110100a	Wheat bread	293	491	163	947
Pasta	110100b	Dry wheat pasta	337	565	187	1,089
Cookies	110100	Wheat flour*	337	565	187	1,089
Rice	100630	White rice	1,788	503	234	2,525
Beans	70820	Green beans	228	71	111	410
Beef	20110-20230	Average different fresh cuts, refrigerated or frozen	12,780	645	498	13,923
Pork	1221-122b	Average different fresh cuts, refrigerated or frozen	6,123	744	583	7,450
Poultry	10599	Live poultry except poultry weighing more than 185 g	3,530	305	409	4,244
Ham	21011	Ham, shoulders, and cured pieces	6,635	815	632	8,082
Sausages	20610-20630	Edible offal of beef and pork**	6,248	366	284	6,898
Egg	40700	Shell eggs	3,224	294		3,518
Milk	2211-2213	Average of different presentations not concentrated nor sweetened	2,059	523	152	2,734
Cheese	40610-40630	Average fresh cheese, powdered or processed	6,876	593	761	8,229
Fermented milk drinks	40310	Yogurt	1,844	156	136	2,136
Tomato	70200	Fresh tomato	61	85	42	188
Onion	70310	Fresh onion	210	87	83	380
Carrot	70610	Fresh carrot	58	49	40	147
Potato	70190	Fresh potatoes	138	112	13	263
Banana	80300	Fresh banana	321	181	35	537
Apple	80810	Fresh apple	977	400	54	1,431
Orange	80510	Fresh orange	524	217	79	820
Bottled water	–	Water content and PET bottle	2	2	13	17
Soda	–	Hypothetical carbonated beverage	348	18	27	393

Source Data from Mekonnen/Hoekstra (2010a, b) and Ercin et al. (2011)

Corn: 50 % grain corn and 50 % cornflour

Soft drink: data from Ercin et al. (2011), except sugar, which was used from those of Mexico (sugarcane, sugar beet, and fructose)

Bottled water: contained water and PET bottle

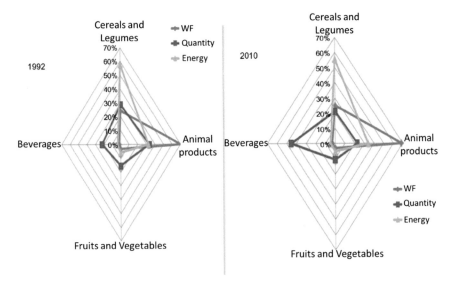

Fig. 9.1 Consumption pattern: amounts, energy contribution, and water footprint by food groups. *Source* Based on data from ENIGH (1992–2010), Mekonnen/Hoekstra (2010a, b), Ercin et al. (2011), and NUTRIPAC Program

Between 1992 and 2010, Mexican population grew at an annual rate of 1.6 %, accumulating a growth of 29 % in the period. The population grew from 86.9 million (M) people in 1992 to 112.3 million in 2010.[6] This increase caused the water required for food to change from 60.6 to 76.12 billion m^3 per year.

What would be the VW requirement associated with food if there has not been a change in dietary pattern? For a possible answer, the 1992 consumption pattern was multiplied by the population in 2010; with the 1992 dietary pattern, national CWRF in 2010 would be 78.4 billion m^3, suggesting that the change in the dietary pattern saved about two billion m^3 of water, water amount equivalent to that contained in the consumption of bread and rice in 1 year.

Figure 9.1 shows the relationship between quantities consumed, VW and energy contribution of the Mexican dietary pattern in 1992 and 2010. The food is organized into four groups: fruits and vegetables and the beverage group are characterized by a significant contribution in quantity, but no significant contribution in water and calories. The cereal group is calorie-intensive with medium contribution to the CWRF and quantity. Finally, animal products are water-intensive.

---

[6]The 1992 population datum is an estimate from the 1990 Census; the datum for 2010 is a result of the census of that year (INEGI).

9.2.2.2 Scenario II

Mexico takes a decisive step toward trade liberalization with the signing of NAFTA in 1994. From 1980 to 2010, the degree of openness of the agri-food sector increases and exceeds the degree of openness of the economy as a whole. Therefore, the dependency ratio[7] of the main food consumed in Mexico increased. Only three foods did not increase their import: egg, milk, and onion. In 2010, rice, wheat, sausages, and pork had the highest dependency ratios (FAO Stat). The increment in consumption of food produced elsewhere was the result of different production processes, qualitative and quantitative, thereby losing food sovereignty.

One objective of the agricultural chapter of NAFTA was to 'integrate' the food market of Mexico to that of the United States, an objective that was fulfilled. 80 % of agri-food imports come from the NAFTA region, particularly from the United States. It is possible to estimate the CWRF with the VW flow of the food imports from the United States, using the WFP of that country utilized by Mekonnen/ Hoekstra (2010a, b).

Compared with Scenario I of sovereignty, in this new scenario the CWRF decreased 2.7 %, from 659.6 m^3 in 1992 to 641.6 m^3 in 2010, a reduction that is due to the changes in dietary patterns already mentioned and the increment in imports from the United States. When considering only data from 2010 and comparing the two scenarios (sovereignty and free trade), it is found that the increase in food imports caused a significant decline in VW content of tortillas, rice, beef, pork and poultry, ham, sausages, cheese, and apples. For wheat products, increased imports caused an increment in virtual water content.

The national CWRF in 1992 was 57 billion m^3 and in 2010, 72 billion m^3. In Scenario I, there was a 'saving' of about 2 billion m^3 of water associated with the change in dietary patterns. Now, it is possible to estimate the effect of trade openness by comparing data from Scenarios I and II. In 1992, the national CWRF, with imports, is lower in 3 billion m^3 than the CWRF of Scenario I of sovereignty; in 2010, this difference is of 4 billion m^3.

A second hypothetical situation (Scenario III) estimates the effect of NAFTA on national CWRF. The CWRF of consumption in 2010 is calculated with the degree of food dependence of the imports in 1992. If the share of imports with the level of 1992 had been kept, the CWRF of Mexico in 2010 would have been 76 billion m^3, this is 3.9 billion m^3 more than with the actual openness and increased imports resulting from the North American Free Trade Agreement.

---

[7]Degree of dependency: imports divided by internal production.

9.3   Conclusions

Food is a fundamental condition of any society; it must be met everyday to ensure the life of individuals. However, societies still fail to ensure a diet matching human dignity and flourishing for the entire population. The threat of a food crisis is still present and the efforts of international organizations to reduce poverty, inequality, and malnutrition, in all its forms, have been insufficient.

The current global food crisis is not an isolated phenomenon; it relates to health and environmental crises. The global agri-food system has caused the impoverishment of rural workers, the double burden of diseases and destruction, pollution and deterioration of ecological resources.

This study considers the proposal to make visible the role of water in the process of social reproduction and, in particular, in food sovereignty. It seeks to understand the interrelationship between water and food in the complex processes of changing dietary patterns and the way the supply of these foods is organized.

Results show that with the change in food consumption and trade openness, Mexico has 'won' in environmental terms by reducing virtual water requirements associated with food. However, when considering the other two dimensions of food sovereignty, it can be seen that this progress in the sustainability of the Mexican food system is accompanied by a deterioration in food that has caused a serious incidence in chronic-degenerative diseases. It has also caused a serious deterioration in the right to local food production and sovereign determination of food policy. The large increase in dependence on food imports has placed Mexico in a very fragile situation to changes in the global food market, particularly, to the increase and volatility of international food prices.

In the current environmental crisis, it is necessary to understand and appreciate not only nature but the complexity of social processes, their dynamics, and multiple powers that shape it.

The way the global agri-food system is organized has been subject to criticism because its mechanisms reproduce and perpetuate the inability of societies to guarantee the right to food, with dignity and justice. These mechanisms may be associated with the power of large agri-food corporations that are behind the loss of sovereignty and deterioration of food. These important economic and political powers hinder changes in the way food is produced, distributed, and consumed to eradicate hunger and improve health.

References

* indicates internet link (URL) has not been working any longer on 8 February 2016.

Araghi, Farshad. "Peasants, Globalization, and Dispossession: A World Historical Perspective", paper presented at the Annual meeting of the American Sociological Association, Hilton San Francisco, San Francisco, Aug 08, 2009.

Bermúdez, Odilia and Katherine Tucker. "Trends in dietary patterns of Latin American populations", *Cad. Saúde Pública* 19(1), (2003):87–99.

Cepede Michel and Maurice Lengelle. *Économie Alimentaire du Globe, essai d Interpretation*, Paris: Librairie De Medicis, Editions M.Th. Genin, 1953.

Clark, Sara, Corinna Hawkes, Sophia Murphy, Karen Hansen-Kuhn and David Wallinga. "Exporting obesity: US farm and tradepolicy and the transformation of the Mexican consumer foodenvironment", *International journal of Occupational and Environmental Health*, vol. 18, no. 1, (2012): 53–65.

Cuéllar, Jose. Programa de seguridad alimentaria: experiencias en México y otros países, México: CEPAL-México, 2011.

*EPC-El Poder del Consumidor. *El fracaso al combate de la epidemia de diabetes y obesidad en México*, http://www.elpoderdelconsumidor.org/fabricaweb/wp-content/uploads/Muertos-Calderón-05_10_12.pdf, Accessed May 20, 2012.

Ercin, Ertug, Maite Martinez and Arjen Hoekstra. "Corporate Water Footprint Accounting and Impact Assessment: The Case of the Water Footprint of a Sugar-Containing Carbonated Beverage", *Water Resour Manage 25* (2011):721–741.

FAO (2012), Water and food security, FAO-UNWATER.

*Forum for Food Sovereignty. *Statement of the NGO/CSO Forum for Food Sovereignty*, http://www.redes.org.uy/wp-content/uploads/2008/09/declaracion_final_del_foro_de_las_ongs_y_movimientos_sociales_en_roma.pdf, Accessed March 4, 2012.

Friedmann, Harriet. "Distance and durability: Shaky foundations of the World Food Economy" in *The global restructuring of agri-food systems*, edited by Philip McMichael, 258–276. USA: Cornell University Press, 1994.

GRAIN. "Soberanía Alimentaria y sistema alimentario mundial", *Biodiversity* 47, 2006. http://www.grain.org/es/article/entries/1086-soberania-alimentaria-y-sistema-alimentario-mundial, Accessed June 16, 2012.

GRAIN. *Food and climate change: the forgotten link* http://www.grain.org/entries?utf8=&q=food, Accessed June 16, 2012.

GRAIN. *The international food system and the climate crisis,* http://www.grain.org/entries?utf8=&q=food, Accessed June 16, 2012.

Hoekstra, Arjen and Ashok Chapagain. "Water footprints of nations: Water use by people as function of their consumption pattern", *Water Resources Management* 21. (2007): 35–48.

INEGI. *Encuesta nacional de ingresos y gastos de los hogares,* México: INEGI, 2012.

Liu, Junguo and Hubert, Savenije. "Food consumption patterns and their effect on water requirement in China", *Hydrology and Earth System Sciences* 12, (2008): 887–898.

Malassis Louis and Gérard Ghersi. *Initiation a l'economie agro-alimentaire*, France: Hatier-Universites Francophones, 1992.

Mekonnen, Mesfin and Arjen Hoekstra. "The green, blue and gray water footprint of farm animals and the animal products", In *Value of Water Research Report Series* 48, Delft: UNESCO-IHE, 2010.

Mekonnen, Mesfin and Arjen Hoekstra. "The green, blue and grey water footprint of crops and derived crop products." In *Value of Water Research Report Series* 47. Delft: UNESCO-IHE, 2010.

NYELENI. *Declaration of NYELENI, Forum for Food Sovereignty*, http://www.nyeleni.org/, Accessed May 27, 2012.

Ocampo, Nasheli and Gonzalo Flores. "Mercado mundial de medios de subsistencia: producción, consumo y circulación de alimentos estratégicos, 1960–1990", Bach diss, Faculty of Economics, UNAM, 1994.

Pérez, Rosario. *El tratado de libre comercio de América del Norte y la ganadería mexicana*, México: UNAM, 1997.

Ramírez, Ivonne et al. "Fruit and vegetable intake in the Mexican population: Results from the Mexican National Health and Nutrition Survey 2006", *Salud Pública de México*, Supplement 4, (2009): 74–85.

Rivera Juán et al. "Consumo de bebidas para una vida saludable: recomendaciones para la población mexicana", *Salud Pública de México*, vol. 50, no. 2, March–April (2008): 172–194.
WHO-FAO. *Diet, Nutrition and the Prevention of Chronic Diseases*, WHO Technical report series Geneva: WHO/FAO, 2003.
*World Food Conference, *Universal Declaration on the Eradication of Hunger and Malnutrition 1974*, http://www2.ohchr.org/spanish/law/malnutricion.htm, Accessed October 30, 2012.

Chapter 10
Hydrological Stress and Pressures on Water Availability

Patricia Phumpiu-Chang

Abstract As a scarce resource and a public commodity, water and its management generate social, political, and economic problems. The purpose of this chapter is to illustrate what the Mexican conditions regarding water availability and the allocated usage and the conflicts that water stress can lead to.

Keywords Water availability · Water stress and water conflicts

10.1 How Much Water Does Mexico Have and Where Is It?

The objective of this section is to introduce the basis for the analysis of pressures over water usage and to understand the circumstances that determine water availability.

There are two types of water sources: renewable and nonrenewable. Renewable water sources are those whose burden/load is more or less equal to the flow they have annually. Nonrenewable sources are those whose burden/load is significantly lower than the water load they contain under normal conditions (Foster/Loucks 2006). This essay focuses on renewable resources, especially ones that are directly influenced by the cycles of annual precipitation. Nonrenewable sources are not considered in this work because they are exploitable only once and cannot be counted on continuously for the volume they possess.

On average, Mexico receives 1,489 billion cubic meters (Km^3) in annual precipitation.[1] However, it is estimated that 73 % of this rainfall is returned to the hydrological cycle, i.e., the rest of the water from the rain 27 % will be available for consumption (Conagua 2011), approximately 460 Km^3 not 402 Km^3. The amount of water availability varies during the year, due to a significant variation in precipitation depending on the season of the year. On average, from May to October, rainfall registers 86.2 % of the annual rainfall (Conagua 2013), indicating that in the

---

[1]Data from 1971 is used to calculate the annual mean incidence in annual precipitation, which includes natural phenomena such as hurricanes (Conagu 2011).

© The Author(s) 2016
R.H. Pérez-Espejo et al. (eds.), *Water, Food and Welfare*,
SpringerBriefs in Environment, Security, Development and Peace 23,
DOI 10.1007/978-3-319-28824-6_10

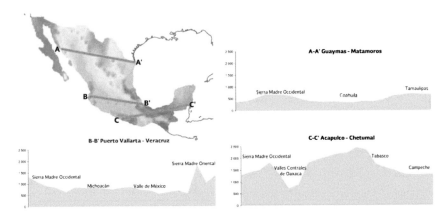

Fig. 10.1 Rainfall Distribution in México. *Source* Conagua (2013)

rest of the year, a period of relative drought is experienced. Notwithstanding, rainfall is not the only relevant source for water availability.

Mexico measures 1.9 million km^2 and it is one of the few countries where eight different types of ecosystems can be found (INECC 2009). Consequently, there are different types of climates with various patterns of rain, and these climate trends indicate that in the north of the country water *from precipitation* is scarce, while in the south *water from precipitation* is abundant (Fig. 10.1).

As a consequence of these climate trends, the main locations with higher availability for water can be determined: in the north of Mexico there are sources with deficit and few sources with water availability due to exploitation (Fig. 10.2), while in the south the resource supply is abundant.[2] Due to overexploitation, México now has rules in an attempt to control water extraction; Fig. 10.2 shows in red the places where restrictions are applied.

10.2 How Is Water Used in Mexico and Where?

Water usage can be consumptive and nonconsumptive. Consumptive usage refers to the extraction of the resource for exploitation in an area different from the source. An example of consumptive water usage is agriculture. Nonconsumptive water usage refers to returning back to the source all the amount of water taken *for human use*. An example of nonconsumptive water usage is the generation of electricity through turbines (Zarco/Mazari 2006). This essay analyzes the consumptive uses only because they create a significant level of stress on water resources.

---

[2]From all water available in Mexico (460 billion liters), Mexico has in concessions 80.5 billion liters (Conagua 2011).

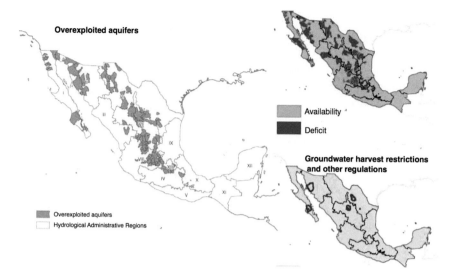

Fig. 10.2 Aquifer sources and overexploitation in México. *Source* Conagua (2013)

There are four consumptive water uses in Mexico (see Chap. 4 of this book). These four types of usage differ only slightly from the *average* global demand levels where food production represents 70 % (AQUASTAT 2013) *of all water consumed*, while in Mexico, food production represents 76.7 % (Fig. 10.3), not such a higher percentage from global levels. Water is a relevant component in the agriculture sector worldwide, and Mexico is not the exception.

Water consumption has a geographic dimension, in the same way as water availability; it is vital to understand how water consumption is distributed. Accordingly, we focus on the most important application: agriculture. Normally, one would assume that the greatest concentration of agricultural activity would locate in

Fig. 10.3 Water use in México. *Source* Conagua (2011)

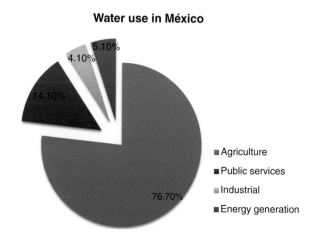

Water use in México

- Agriculture
- Public services
- Industrial
- Energy generation

76.70%
14.10%
4.10%
5.10%

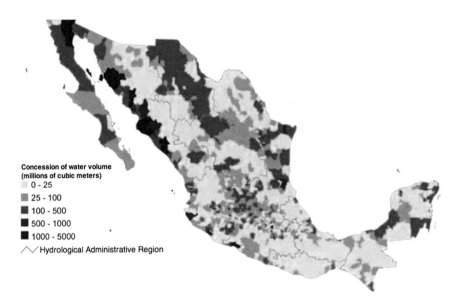

Fig. 10.4 Water demand in México. *Source* Conagua (2011)

areas with greater precipitation; however, this is not the case, since Chihuahua, Sinaloa, and Sonora, in the north of the country, *are given* the greatest volume of water concessions (Fig. 10.4) by the federal government representing 27.1 % (21,852.3 million liters) of the total *national* water consumption and 31.8 % (19,640.3 million liters) of the *national* total *of water concessions* for agricultural use.

10.3   How Is Water Stressed in Mexico?

The imbalance from water supply and demand should have the answer to the question of not "how much of the water resource is available?", but instead "how much of water resource is needed for the market to meet the demand?" Let us understand how many people need water.

Mexico has a population of 112 million people (World Bank 2013), and on average, each Mexican consumes 1965 l a day.[3] The United Nations Food and Agriculture Organization (FAO) estimates that a person consumes 2000–5000 l of water a day (AQUASTAT 2013). Compared to the global *standard of* demand and supply, water resources distribution in Mexico seems to indicate some level of

---

[3]This amount was calculated as follows: Total daily consume = Annual Grant (80.5 billion liters)/ Total population (112 million people)/Days of the year (365) consumption.

scarcity—3 % below the estimated global minimum—which does not mean that there is indeed a shortage.

The diet of the average Mexican is based mainly on the consumption of grains and legumes, which are not water-intensive products. Meanwhile meat, a water-intensive product to produce, is consumed in smaller proportions (Martínez/Villezca 2003). Consequently, we could assume that there is not a significant level of stress on water consumption because the demand on consumption is satisfied, but this is not entirely true.

Geographical distribution plays a relevant role in water security, and there is a defined differentiation of population among the 13 Hydrological Administrative Regions (RHA). While five regions exceed ten million inhabitants, seven regions do not exceed five million inhabitants, indicating the existence of inhabitants' clusters. Accordingly, it can be inferred that there is a high level of pressure on water in these five regions, while in the remaining regions water demand is significantly less (Table 10.1).

The wide variations in water volumes for concessions to each RHA, the pressure level in water use, and the consumption of water per capita variations are discussed below in relation to productive activities.

Table 10.1 Level of water stress in the RHA

Hydrological administrative region	Concession of volume of water (m³)	Renewable water volume (m³)	Water pressure (%)	Water consumption per capita per day
Peninsula de Baja California	3420000000.00	4667000000.00	73.3	2360.17
Xoroeste	7703000000.00	8499000000.00	90.6	8170.39
Padfico Norte	10411000000.00	25630000000.00	40.6	6828.65
Balsas	10704000000.00	21680000000.00	49.4	2668.43
Padfico Sur	1363000000.00	32824000000.00	4.2	782.86
Rio Bravo	9243000000.00	12163000000.00	76.0	2241.99
Cuencas Centrales del Norte	3846000000.00	7898000000.00	48.7	2480.46
Lerma-Santiago-Pacifico	14479000000.00	34533000000.00	41.9	1776.78
Golfo Norte	4854000000.00	25564000000.00	19.0	2669.34
Golfo Centro	4973000000.00	95866000000.00	5.2	1360.83
Frontera Sur	2203000000.00	157754000000.00	1.4	854.90
Peninsula de Yucatan	2731000000.00	29645000000.00	9.2	1823.59
Aguas del V'aLle de Mexico	4658000000.00	3513000000.00	132.6	584.99
Total	80588000000.00	460236000000.00	17.5	1965.52

Source Conagua (2011)

In the "Noreste" region II, Sonora produces 4.96 million tons of food a year. It is important to highlight the type of production, which is 1.8 million tons of wheat and 316,000 or 3.16 million tons of meat[4].

The sum of food production in the three regions of the "Pacífico Norte," "Balsas," and "Lerma-Santiago-Pacífico" regions exceeds 280 million tons. For these three cases, one should take into account that the level of pressure on the resource does not exceed 50 %.

The "Rio Bravo" region features a low rainfall rate and much of its *water* supply depends on the Rio Bravo, whose aquifers report high levels of overexploitation (Conagua 2013). *Agricultural water consumption is distributed as follows: Chihuahua 90 %, Coahuila 83 %, and Nuevo León 71 %, have a percentage of total water concession.*

In extreme opposed cases, RHA "Pacífico Sur," "Golfo Centro," "Frontera Sur," and "Península de Yucatán" have a low pressure level over the resource, indicating a lower demand in these regions. However, according to Conagua (2011), the *per capita* water consumption is much lower than the national average of 1965 l, with the exception of "Península de Yucatán" area, where consumption is 1823 l/day.

From the above it can be said that Mexico is a country that does not have an average level of significant pressure on the resource, but in specific circumstances the pressure can be very high. Of the 112 million people, 81.4 million people live in areas with a level of pressure on the resource above 40 %. In the long term, in these areas, this pressure will seriously erode the ability of the aquifers to regenerate completely without damaging the environment or water provided (Foster/Loucks 2006). Conagua (2010) explains the reason why Mexico as a country seems to have no relevant pressure over water on the average is that agriculture depends directly on the rain and this water is not accounted for in the concessions granted by the state.

The main problem is that the demand *for water is increasing.* The Mexican population is experiencing a population growth of over 1 % per year, which means that each year more than one million people join the population, and all will have the *same* right to consume water as *did the previous generations* (World Bank 2013), worsening existing conflicts over water.

10.4 What Conflicts Emerge from the Water Stress Level?

The situations described generate, inevitably, disputes over the right to water. Questions such as "who should have priority when distributing the water?" and "how much water is really needed?" tend to create tensions among different actors interested in the resource.

---

[4]Wheat demand 1300 l of water to produce a kilo. Beef demand approximately 15,000 l of water to produce a kilo (Mekonnen/Hoekstra 2011).

10.4.1 Water in the Colorado River

In most cases, water tensions occur among actors who share the same nationality. However, there are cases of shared international waters denoting more complex situations; one of the most important examples is the Colorado River, a flow that is born *in the western United States* and empties into the Gulf of California in Mexico, granting rights to both countries for exploitation of its waters (Hundley 2000).

The relationship between Mexico and the United States is long and complicated and it is a *sui generis* case compared to traditional diplomacy. Throughout history, there have been conflicts that have led to war and the transfer of territory from one country to another.

Currently, the Colorado River basin generates little tension between the two nations because there is an agreement in force since 1944, in which a guaranteed amount of water—*1.85 billion cubic meters—is provided by the United States* to Mexico (SRE 1944). This agreement does not specify the quality of the water concessions.

The agreement opened the possibility for building infrastructure for river water management; however, these constructions *in the United States* polluted the river *downstream* with sediment that affected water salinity. With high levels of salinity, the quality of the water flow was not adequate for agriculture usage *in Mexico*: several *seasons of* crops were lost threatening food security (Hundley 2000).

In 1907, Mexico asked for more than 5 billion cubic meters of water from the United States per year. The negotiations were interrupted by the Mexican Revolution and by 1944, when it was signed the joint agreement of shared governance was made in the context of the Second World War. By this time, diplomatic relations between the two countries had intensified significantly and the joint agreement gave United States the assurance of voluntary cooperation of the Mexican state in the future.

The water salinity issue, which brought back the tensions between the two countries, was due to the different usages of the water by the two countries. Mining and power generation in the United States contaminated water with sediment that affected agricultural usage south of the border that required a low level of salinity. Mexico had a secure amount of water, even if it was contaminated.

It was not until 29 years later that a final agreement determined the acceptable salinity of water delivered to Mexico, a maximum of 150 parts per million. However, the salinity of water delivered has increased over the years—leaving the U.S. at 153 parts per million in 2011 (IBWC 2013).

At present, the agreement signed in 1944 represents a problem because the political and economic conditions in which it was signed no longer exist: the Mexican population has multiplied by 4 since 1944; according to INEGI (2013a, b) in 1944 there were only 25 million people in Mexican national territory.

In international law, all documents can be declared invalid if the conditions in which they signed no longer exist (Ahlf 2004). The flow of the Colorado River is one of the most legislated worldwide, where national laws and international treaties determine water allocation (Ruiz 2011). Such stringent legislation is ineffective because of the increasing growth of the population and thus the increasing demand for the resource. Even if the population does not increase significantly, the water demand will continue to grow due to productive activities causing stress on the allocated amount of water.

The Hydrological Administrative Region (RHA acronym in Spanish) at the mouth of the Colorado River has a demand of 11 billion cubic meters (Conagua 2011); the high water demand has generated significant exploitation of the aquifers and has significantly contaminated them. If the rate of consumption continues to increase, the availability of water in the area will decrease with consequent increase in the level of stress on the resource: the water balance in the flow of the Colorado River will deteriorate *further*.

The United States strongly protects power generation and mining in the Colorado River area, it would not be willing to sacrifice for the cost of these activities. Consequently, Mexico has no priorities over American priorities: Mexico has no bargaining power to restore the water balance in its favor. We are talking of only 2 % of Mexico's water supply and a full quarter of the total consumed in the area. Achieving a satisfactory and definitive agreement took Mexican diplomacy 70 years; we can observe the difficulty of the situation.

The RHA "Noroeste" is particularly vulnerable to water availability. With a demand for 8170 l of water per capita per day, the priority for water usage in this area is food production. Limiting the amount of water in an area of food production in a country with a growing population at a rate of 1 % per year (World Bank 2013) strongly compromises food security. However, more water from the northern border cannot be obtained because of the delicate diplomatic equilibrium that will break if there is a renegotiation of the existing treaties. In this case, it is necessary to reduce the demand for maintaining a sustainable use, but *such an action* will strongly affect economic activity.

Trying to decrease water allocation by increasing the water fees is counterproductive due to the nature of the demand. If water pressure could be decreased by lack of sanitation and industry, it would be an effective measure; but food production has an inelastic demand, so increasing water fees would only serve to reduce profit margins *with added costs onto customers* by an increase in the value of inputs.

10.4.2 Water in Mexico City and Its Metropolitan Area

The case of Mexico City is different from the rest of the country because it is located on a major aquifer, whose the population exploits to meet their needs. However, water availability is limited by natural processes, i.e., by dependence upon rain cycles that regenerate aquifers.

Mexico DF has a population of 29 million people, and it is one of the most densely populated cities in the world. The RHA "Aguas del Valle de México" is a place where the level of pressure on the resource is greater than 100 %. This means that neither accumulating nor distributing all the water that naturally flows into the area is sufficient to meet the demand of the population; the questions that arise are what is the water usage and from where is the demanded missing water obtained to cover the shortfall?

The first question is related to the predominant economic activities in the area. The use of water for sanitation is a constant in any populated area; however, it only represents 14 % of national water demand. The remaining water usage has a direct relationship with the predominant economic activities in each area, industrial and agricultural use type demand.

The structure of economic activity in the Federal District is: 84.4 % tertiary activities[5]; 15.53 % secondary activities[6]; and 0.06 % primary activities[7] (INEGI 2013a, b). A fact that should be taken into account is that Mexico City Federal District (DF) also comprises areas with agricultural production (Conagua 2010). This distribution of the economic activity indicates that food production is not a demand for relevant resources to the area. Consequently, water allocation for agriculture or livestock is not a priority, which frees the area with a need for much greater use. Thus, the average water used per capita is lower than the national average.

Mexico DF belongs to an RHA that has an average water demand of 585 l per person per day, and the main water usage is allocated for public service supply and industry. The national average for these uses is 458 l per day per person, but the value of this average is higher in DF because 99.5 % of the population lives in urban areas (INEGI 2013a, b). Urban areas have a greater demand for water for sanitation than rural areas; however, this disparity does not correspond to a greater need, but to a distribution problem. The development of cities allows people better access to water, practically fulfilling their needs fully, contrary to rural areas where the main access to water for sanitation is not from a sewer pipeline system, but directly from nearby aquifers (WHO 2007). It is not that the population in rural areas needs less water than those living in urban areas, but the resources needed to meet their demand is not accounted for within the state concession system because water is not considered a public good, access to which is guaranteed by the government.

The remaining water demand is obtained from a nearby water system. The Cutzamala system is used to supply the Federal District with 29 % of the total demand. This system extends for more than 100 km away from the area where the water is obtained by means of a concession outside the Federal District. As shown in the analysis of the Colorado River, allowances based on sources outside the use zone generate conflict between the different actors involved, and this case is no exception. When it comes to a single country, federal laws apply and under these

---

[5]Commerce, restaurants, hotels, transport, financing services, education services, medical services, real state services, and governmental activities.

[6]Mining, construction, electricity, water services, gas services, and manufacturing industry.

[7]Agriculture, livestock, forestry, fishing, and hunting.

laws the government has the ability to expropriate any land with resources, which is important for national security (ITESM 2013). Without constraints, the State can designate which resources should be used in which area, including water. However, the conflicts that occur in these cases do not represent a true infeasibility in sharing resources. The disparities presented are derived mainly from a sense of ownership over water (Sainz/Becerra 2013). Conflicts over property rights are normally mitigated by the Basin Councils, bodies that exist to mediate between the various stakeholders in water resources and ensure equitable distribution as much as possible. The Cutzamala system is regulated by federal law and Basin Councils (FCEA 2013) function to prevent the escalation of conflicts.

Water usage by one RHA of water from another RHA raises pressure on the resource in the source from where it is obtained. However, the marginal cost is not high if the water is obtained from an area where demand per capita is lower than the water-deficient RHA. To obtain one liter of water in a region where water is not available (RHA "Aguas del Valle de Mexico"), and to obtain one liter of water in a region where there is an availability of approximately 50 % (RHA "Lerma-Santiago-Pacific") is not the same thing.

10.5   Conclusions

The availability of water in Mexico is not a situation that can be understood in a framework of generalities. Analysis of the factors that cause water stress and pressures is needed from the root causes, for the specific case. Because of its wide territorial extension, the country has diverse ecosystems that behave differently in terms of water availability. Considering that the population distribution is not uniform, we have to take into account that water demand is equally uneven. The analysis of water availability by basin allows us to have an approach more grounded in the Mexican reality. In this country, 80 million people live in areas where the level of pressure on the resource is over 40 %, and 25 % of this population lives in an area with a higher pressure level of 100 %. There is water in Mexico; governance, management, and awareness are the answers for supply and demand to each location.

References

* indicates internet link (URL) has not been working any longer on 8 February 2016.

Banco Mundial. *World Bank Data Base,* http://datos.bancomundial.org/pais/mexico, Accessed November 5, 2013.
Bureau of Reclamation. *Hoover Dam Frequently Asked Questions and Answers.* http://www.usbr.gov/lc/hooverdam/faqs/powerfaq.html, Accessed November 15, 2013.
Carreño, Alberto. *La diplomacia extraordinaria entre México y Estados Unidos.* México: Jus, 1961.

Castro, José Luis et al., "La frontera México-Estados Unidos" in *Retos de la Investigación del Agua en México,* edited by Úrsula Oswald, 483–492. México: UNAM, 2011.

CILA. *Río Colorado,* http://www.sre.gob.mx/cilanorte/index.php/rio-colorado, Accessed November 16, 2013.

CONAGUA. *Estadísticas agrícolas de los distritos de riego.* México: SEMARNAT, 2010.

CONAGUA. *Estadísticas del agua en México.* México: SEMARNAT, 2011.

*CONAGUA. *Atlas digital del agua México 2012.* http://www.conagua.gob.mx/atlas/ciclo14, Accessed November 2, 2013.

FAO, AQUASTAT. *Water Uses,* http://www.fao.org/nr/water/aquastat/water_use/index.stm, Accessed 5 de November de 2013.

*FCEA. *Comision Especial: Cuenca del Sistema Cutzamala,* Centro Virtual de Información del Agua, http://www.agua.org.mx/h2o/index.php?option=com_content&view=article&id=11960: comision-especial-cuenca-del-sistema-cutzamala&catid=1221:camara-de-diputados, Accessed November 25 de November de 2013.

Foster, Stephen and Daniel P. Loucks. *Non Renewable Groundwater Resources.* París: UNESCO, 2006.

Hundley, Norris. *Las aguas divididas: un siglo de controversia entre México y Estados Unidos.* México: Prisma Editorial, 2000.

INECC. *Los ecosistemas de México.* http://www.inecc.gob.mx/con-eco-ch/382-hc-ecosistemas-mexico, Accessed August 26, 2009.

INEGI. *Anuario de estadísticas por entidad federativa.* México: INEGI, 2012.

INEGI. *Actividades economicas del Distrito Federal.* http://cuentame.inegi.org.mx/monografias/informacion/df/economia/default.aspx?tema=me&e=09, Accessed November 15, 2013.

INEGI. *Censo de 1950,* http://www.inegi.org.mx/prod_serv/contenidos/cspanol/bvinegi/productos/integracion/pais/historicas2/cienanos/EUMCIENII.pdf, Accessed November 17, 2013.

*ITESM. *Ley de expropiación,* http://www.cem.itesm.mx/derecho/nlegislacion/federal/36/1.htm, Accessed November 18, 2013.

Martínez, Irma. and Pedro A. Villezca. "La alimentación en México: un estudio a partir de la *Encuesta nacional de ingresos y gastos de los hogares",* *Revista de Información y Análisis,* (2003): 26–37.

Mekonnen, Mesfin and Arjen Hoekstra. "The green, blue and grey water footprint of crops and derived crop products" *Hydrology and Earth Systems Science,* (2011): 1577–1600.

Ohlsson, Leif. "The Role of Water and the Origins of Conflict", in *Hydropolitics* edited by Leif Ohlsson, 1–28. UK: Zed Books, 1995.

OMS. *La meta de los ODM relativa al agua potable y el saneamiento: el reto del decenio para zonas urbanas y rurales.* Suiza: ONU, 2007.

Ortiz, Loretta, *Derecho internacional público.* México: Oxford University Press, 2004.

SAGARPA. *Estudio de gran visión y factibilidad económica y financiera para el desarrollo de infraestructura de almacenamiento y distribucion de granos y oleginosas,* México: SAGARPA, 2009.

Sainz, Jaime, Mariana, Becerra and Carlos Muñoz. *Los conflictos por agua en México,* http://www.inecc.gob.mx/descargas/dgipea/conf_agua_mex.pdf, Accessed November 12, 2013.

SRE. *Tratado entre el Gobierno de los Estados Unidos Mexicanos y el Gobierno de los Estados Unidos de América de la Distribución de las Aguas Internacionales de los Rios Colorado, Tijuana y Bravo.* Washington: 1944.

Zarco, Alba E. and Marisa Mazari. "Usos consuntivos de agua superficial y subterranea", en *Atlas de la Cuenca Lerma-Chapala,* edited by Helena Cotler, Marisa Mazari y José de Anda, México: Instituto Nacional de Ecología, 2006.

Chapter 11
Problems Associated with Groundwater Management

Eugenio Gómez-Reyes

Abstract An important component of water supply of the country is the supply from groundwater sources. They correspond to the use of blue water and they complement the surface water supply. Unlike surface supply sources, for which there is abundant information and analysis, groundwater resources are an area of water management that needs institutional strengthening by increasing their knowledge. Despite this, the diagnoses of this type of resources provide clues that indicate a growing overexploitation and competition, which show the results of institutional arrangements that have promoted the use of water in the country.

Keyword Underground water and management problems

11.1 Groundwater Resources

For the purpose of groundwater management, the country has been divided into 653 aquifers, whose official names were published in the Official Government Gazette of December 5, 2001 (Conagua 2011). The number of aquifers by Hydrological-Administrative Regions (HAR) is shown in Table 11.1, which also presents the extraction, recharge, and availability conditions of the volume of groundwater resource. The statistics in Table 11.1 indicate that the country has an availability of 63 % (1636.130 m^3/s) of its underground resources with respect to the recharge volume. This availability is evident especially in the south, in the XI Southern Border and XII Yucatan Peninsula HAR, where groundwater extraction does not reach 10.5 % of the aquifer recharge.

In the southern and southeastern HAR (IV Balsas, V Southern Pacific and X Central Gulf), groundwater resources availability with respect to discharge is greater than 50 % (Fig. 11.1). On the other hand, the northern HAR, I Baja California Peninsula and VII Central Basins of the North, have exploitation conditions (extraction is greater than recharge) of 32 and 8 %, respectively. Furthermore, HAR located in the northern and central part of the country, II Northwest, VI Rio Bravo, and XIII Waters of the Valley of Mexico, have conditions close to overexploitation

© The Author(s) 2016
R.H. Pérez-Espejo et al. (eds.), *Water, Food and Welfare*,
SpringerBriefs in Environment, Security, Development and Peace 23,
DOI 10.1007/978-3-319-28824-6_11

Table 11.1 Groundwater resource in Mexico

HAR	Name	Aquifers (No.)	Qunderground		Qcondition		Overexploitation[b] (%)
			Extraction (m³/s)	Recharge (m³/s)	Availability[a] (m³/s)	Recharge %	
I	Baja California Peninsula	87	54.477	41.223	−13.254		32.2
II	Northwest	63	97.317	108.638	11.32	10.4	
III	Northern Pacific	24	41.984	103.596	61.612	59.5	
IV	Balsas	46	57.807	146.594	88.787	60.6	
V	Southern Pacific	35	14.523	64.181	49.658	77.4	
VI	Rio Bravo	100	139.047	168.252	29.205	17.4	
VII	Central Basins of the North	68	82.287	75.85	−6.437		8.5
VIII	Lerma-Santiago-Pacific	127	226.63	256.913	30.283	11.8	
IX	Northern Gulf	40	33.264	42.428	9.164	21.6	
X	Central Gulf	22	29.997	135.084	105.086	77.8	
XI	Southern Border	23	19.375	571.252	551.877	96.6	
XII	Yucatan Peninsula	4	84.316	802.765	718.449	89.5	
XIII	Waters of Valley of Mexico	14	73.789	74.169	0.381	0.5	
Total	Mexican Republic	653	954.814	2590.94	1636.13	63.1	

Source Based on data from Conagua (2011)

[a]Availability = Recharge-Extraction

[b]Exploitation = Extraction/Recharge

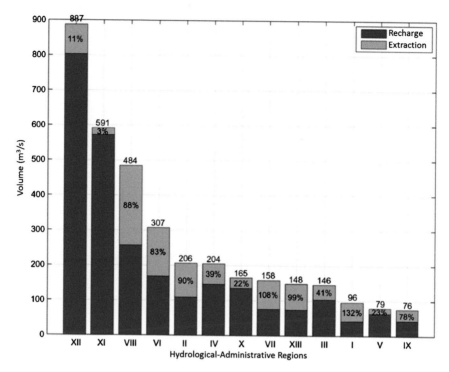

Fig. 11.1 Groundwater volume in HAR, showing the extraction volume percentage compared to the recharge volume. *Source* Based on data from Table 11.1

of groundwater resources, whose recharge available volume scarcely exceeds in 15 % the extraction volume.

Although underground reservoir in the country is substantial (1636.130 m³/s) and only two of the 13 HAR (I and VII) are overexploited, the global balance does not reflect the critical situation of vast arid and semiarid regions, where water balance is negative and underground storage is running out.

Of the 653 conventionally defined aquifers, 100 are subjected to intensive exploitation (Conagua 2011). The most critical cases are in the northwest, north and center of Mexico (Fig. 11.2), particularly in the Lerma River basin (VI HAR), mainly in the states of Guanajuato and Queretaro; in the Laguna Region (VII HAR) in Coahuila, Durango and Aguascalientes; in Chihuahua (VI HAR), Sonora (II HAR), and the Federal District (XIII HAR).

In areas of overexploited aquifers, this situation compromises the sustainable development of all sectors, with serious implications for the national economy, since several of the most important cities are supplied by aquifers. In addition, the intensive use of groundwater has caused a severe ecological impact, causing the disappearance of lakes and wetlands, reduction of the base flow of rivers, and the loss of ecosystems. There also have been other effects such as the deactivation or decrease in the performance of wells; increased costs of deeper water extraction

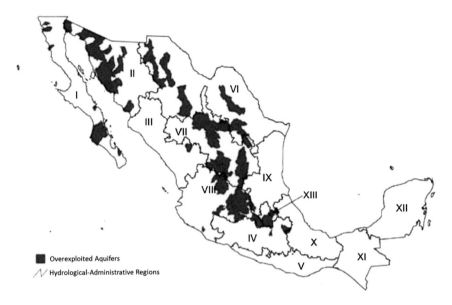

Fig. 11.2 Overexploited aquifers in Mexico. *Source* Conagua (2011)

wells due to the raising consumption of electricity; land settlement and cracking (subsidence phenomena); aquifer pollution and saline intrusion in coastal aquifers, and a strong competition between users. In many cases, the supply to meet water demand of the cities depends on the release of groundwater previously allocated for other uses, by transfer of rights; this problem worsens by the population and economic growth tendency.

An important problem related to groundwater is the limited knowledge of the most important aquifers; this is acceptable for general purposes of water management, but insufficient to guide the management required to reconcile aquifer preservation and satisfy the growing demands for water (Chávez et al. 2006).

Furthermore, water legislation is insufficient and/or inadequate for effective groundwater management, e.g., most of the closures are inoperative and incompatible with current aquifer operating conditions.

11.2 Groundwater Concession

As mentioned in the chapter on water use, 37 % (954.814 m^3/sen 2012) of the total volume allocated for offstream uses in the country comes from aquifers. The importance of groundwater is due to the magnitude of the volume used by all sectors (Fig. 11.3). The major use is for agriculture (69 %), followed by public

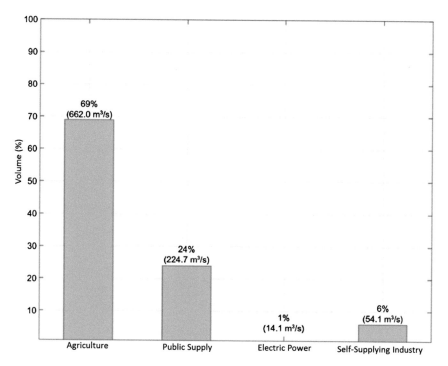

Fig. 11.3 Groundwater volume allocated for offstream uses in Mexico. *Source* Based on data from Repda

supply (24 %); the remaining 7 % is for self-supplying industry (6 %) and electricity generation (1 %). The groundwater for agricultural use (662 m³/s) is the source that provides water to two million hectares, a third of the total irrigated area. For public supply, it supplies 62 % of the volume required by the cities, benefiting about 75 million people. Groundwater also supplies half (51 %) of the industrial facilities.

On the other hand, Fig. 11.4 shows the way in which groundwater has been allocated for offstream uses in the HAR of the country. It can be observed that the VIII Lerma-Santiago-Pacific HAR is the largest consumer of groundwater (227 m³/s), followed by the VI Rio Bravo (139 m³/s) and II Northwest (97 m³/s) HAR. The rest of the HAR consume groundwater gradually, from 84 m³/s in the XII Yucatan Peninsula, up to 15 m³/s in the V Southern Pacific.

It is worth mentioning that most of the groundwater volume allocated for the XIII Waters of the Valley of Mexico HAR (74 m³/s) is for public supply (76 %), and not for agricultural irrigation as in other HAR. The groundwater volume used for agriculture in the VIII Lerma-Santiago-Pacific HAR (168 m³/s) is greater than the one used for this purpose in any of the other Hydrological Management Regions.

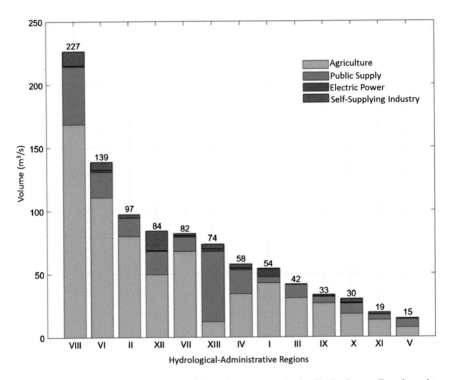

Fig. 11.4 Groundwater volume allocated for offstream uses in the HAR. *Source* Based on data from Repda

11.3 Degree of Groundwater Stress

Similar to the degree of stress exerted on water resources (percentage of water used for offstream uses compared to the renewable water resources), it can be used an index that expresses the degree of groundwater stress caused by agricultural use, i.e., the percentage that represents the groundwater allocated in other offstream uses with respect to agricultural use. Thus, it can be considered that if this percentage is greater than 40 %, there is a strong stress on the agricultural use of groundwater.

Additionally, it can be identified the groundwater use that exerts the highest stress on agricultural use, since each one of the water uses is a partial component of the degree of total stress on the agricultural use of groundwater.

Nationally, the degree of stress on groundwater by agricultural use is 80.5 % on average (Table 11.2), which is considered a high level. This indicates that food security, which depends on irrigation using groundwater, is subject to strong competition between users. In this case, the highest stress (82.5 %) on food security is exerted by Public Supply. The high degree of stress on groundwater resources in the country requires an effective management that includes actions to increase water

Table 11.2 Degree of stress of agriculture on groundwater (percentages)

HAR	Name	PS[a]	SSI[b]	EG[c]	Total	Classification
I	Baja California Peninsula	10.7	1.7	14.7	27.2	Moderate
II	Northwest	18.1	3.5	0	21.5	Moderate
III	Northern Pacific	34.4	2	0	36.4	Moderate
IV	Balsas	55.1	9.4	4.4	69	High
V	Southern Pacific	88.8	8.6	0	97.4	High
VI	Rio Bravo	18.2	5.8	1.7	25.6	Moderate
VII	Central Basins of the North	17	2.9	1.3	21.3	Moderate
VIII	Lerma-Santiago-Pacific	27.2	6.9	0.4	34.6	Moderate
IX	Northern Gulf	19.1	4.7	0.7	24.6	Moderate
X	Central Gulf	47.1	17.9	1.2	66.3	High
XI	Southern Border	30.7	13.7	0	44.4	High
XII	Yucatan Peninsula	37.7	32.3	0.6	70.6	High
XIII	Waters of Valley of Mexico	459	30.8	17.8	507.6	Very high
Average	Mexican Republic	66.4	10.8	3.3	80.5	High

Source Based on data from Conagua (2011)

[a]*PS* Public supply

[b]*SSI* Self-supplying industry

[c]*EG* Electricity generation

availability in aquifers and to promote their preservation, integral, and efficient use and reuse. Meteoric water that has not gone through sources of pollution is a large potential resource to increase aquifer natural recharge. This rainwater can be infiltrated through absorption wells without restrictions as to its quality. On the other hand, despite the high cost of advanced wastewater treatment, this alternative is a considerable potential resource to recharge aquifers artificially because there is a permanent and increasing flow compared to the growth of public demand. In order to prevent groundwater deterioration and damage to public health, especially where there is a risk that the treated wastewater migrates to drinking water intakes, artificial recharge systems should consider subsoil as a natural treatment plant that can be exploited with an appropriate mix of pretreatment–natural treatment–posttreatment, compatible with the recharge method and the intended use for the reclaimed water.

As for the spatial distribution of the degree of stress on agricultural use of groundwater, there is an enormous stress of 508 % in the XIII Waters of the Valley of Mexico HAR (Fig. 11.5). In this HAR, there is one of the most populated urban centers of the planet, Mexico City Metropolitan Area. According to the results of the Census of Population and Housing from 2010, more than 27 million people lived in this area, 24.8 % of the total population, where the greatest contribution to the gross domestic product is generated (just over 20 %). The groundwater allocated for public supply (55.746 m^3/s) and self-supplying industry (3.742 m^3/s), which sustains urban and economic development of this megacity, exceeds almost five times the groundwater allocation volume for agriculture (12.145 m^3/s).

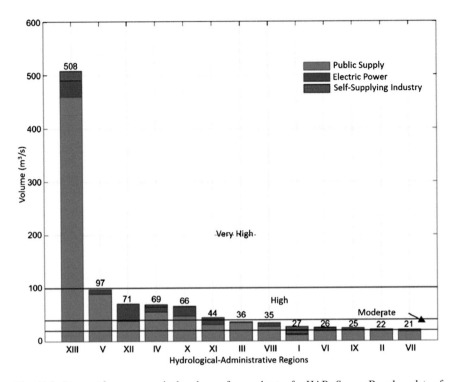

Fig. 11.5 Degree of stress on agricultural use of groundwater for HAR. *Source* Based on data of Table 11.2

In the XIII HAR, it is required an effective groundwater management as a valuable instrument to mitigate the degree of stress on agricultural use. This management should include measures to meet the demand in all sectors as well as actions to increase the natural recharge of the aquifer with rainwater and the artificial recharge with treated wastewater, as suggested for the whole country.

In the urban sector, it is necessary to implement programs to detect leaks and recover lost volume; to tend towards an equitable distribution using metering valves to ensure an appropriate endowment, rather than improve micrometering; and make efficient water use with economic incentives for the installation of devices and water saving systems.

For the industry, it is required the use of treated wastewater for uses where drinking water is not needed; industrial order to encourage the enterprises to generate products that leave a low water footprint and promote the import of those that require large amounts of water in their production processes.

In agriculture, projects for modernization of irrigation are needed to improve efficiency in agricultural water use that includes the change of traditional crops for other more productive and that consume less water, the rehabilitation of agricultural infrastructure, and user training in the application of new technologies.

References

Chávez, Rafael, Francisco Lara and Roberto Sención. "El agua subterránea en México: condición actual y retos para un manejo sostenible" *Boletín Geológico y Minero* 117 (1), (2006):115–126.

CONAGUA. *Estadísticas del agua en México.* México: SEMARNAT, 2011.

Chapter 12
Vulnerability and Climate Change

Hilda R. Dávila-Ibáñez and Roberto M. Constantino-Toto

Abstract This chapter discusses the linkages and the implications of the potential presence of large-scale hydrometeorological events in Mexico and the relevance of the water footprint methodology to strengthen the institutional management capabilities that tend to mitigate the impacts on the welfare of the population at local levels.

Keywords Water · Vulnerability · Climate change

12.1 Introduction

This document analyzes the relations and implications of the potential occurrence of large-scale hydrometeorological events in Mexico and the relevance of the water footprint approach to strengthen the institutional management capacities to mitigate the welfare impacts at local level.

The geographical position of Mexico makes it possible for hydrometeorological major events to occur regularly, directly affecting the population welfare because of the impact caused by floods and droughts on the physical integrity of persons, assets and patrimony of citizens, and the associated damages on public services infrastructure. There are also significant disturbances because of the impact that such events may have on the articulation of local production systems, affecting production capacities, employment, and income.

In recent years, there has been progress in strengthening the government capacity to take care and act if a natural disaster occurs in the country. However, compared to the distributive asymmetries, heterogeneity in the functioning of economic sectors, and stability of the production functions in the use of natural resources, it is necessary to make an effort to reduce vulnerability to potential risks on water (Constantino and Davila 2010).

The exploration of the relationship between vulnerability, hydrometeorological events related to climate change, and water footprint approach is carried out in three sections, in the context of food security. In the first section, the notions of vulnerability and resilience capacity to impacts that alter the stability of socioeconomic

R.H. Pérez-Espejo et al. (eds.), *Water, Food and Welfare*,
SpringerBriefs in Environment, Security, Development and Peace 23,
DOI 10.1007/978-3-319-28824-6_12

systems are addressed. Methodological characteristics that the water footprint approach could contribute to strengthen institutional capacities in management of risks arising from hydrometeorological impacts are established in the second section. In the third section, an approximation of the latent vulnerability to droughts occurring in the federal states is presented, based on cluster analysis.

12.2 Considerations About Vulnerability and Climate Change

The vulnerability of a region, an economic sector, or a country is the result of a complex relationship between the natural and the socioeconomic systems. This can be measured using the capacity that exists in a locality, region, or industry to design and carry out actions that will counter the magnitude of the negative effects associated with threatening events that alter the natural system stability.

The vulnerability is the result of interactions between nature, socioeconomic system, economic growth patterns, use of natural resources, and distributive rules prevailing in a society that affect fundamental welfare levels in the definition of responsiveness and resilience of society to natural disasters. So, regardless of the origin and characteristics of the natural disaster, the responsiveness is largely a social construction.

Regarding the negative impact caused by natural disasters, in general, but that can be applied to phenomena originated by water, Cutter (2008a) established the idea of sustainability as the community ability to provide resilience to negative impacts without requiring external resources. Meanwhile, Perrings (2006) incorporated, from economics, the notion of sustainability as the flexible adaptability against hostile scenarios, without losing functional organizing characteristics.

In the exploration of system vulnerability, not only the conditions of resilience, response, and eventual recovery of the properties that ensure their persistence are important, but also the identification of the origin, magnitude, and duration of events that compromise the stability of the functions of the system is relevant (Vargas 2002; Bitrán 2009). In the case of water-related events, it is also crucial to recognize the fact that eventually the increment of impacts can be caused by their own economic and social practices.

Although the possibility of confronting socially a loss due to a change in nature conditions is inevitable, the local resilience capacity is directly related to the characteristics of population dispersion in the territory, the corresponding welfare levels, and use practices of economic sectors in the areas of impact and its respective distribution.

According to international forecasts, threats associated with climate change are potentially:

- Variations in mean annual temperature, in land and ocean
- Changes in rainfall cycle

Table 12.1 Model with heteroscedasticity correction, using observations 1–18 Dependent variable: Ln VTP$_{AE}$

	Coefficient	Std. Dev.	t-statistics	p-value
const	8.41251	1.63089	5.1582	0.0001***
Ln STA	0.919407	0.176701	5.2032	0.00009***
Statistics based on weighted data:				
Residual sum of squares	41.94016		S.D. regression	1.619031
R-squared	0.628538		Adjusted R-squared	0.605321
F(1, 16)	27.073		p-value (of F)	0.000087
Log-likelihood	−33.15374		Akaike criterion	70.30748
Schwarz criterion	72.08823		Hannan–Quinn crit.	70.55302

Source Own elaboration with data from Conagua

- Variations in the intensity of extreme weather events.

  Issues that would trigger a set of social consequences are:

- Modification of hydrometeorological cycles and effects on water availability
- Loss of biodiversity
- Reduction of food production potential
- Changing patterns of population disease.

The expected impacts of climate change, according to circulation models and scenarios[1] developed by the Intergovernmental Panel on Climate Change (2001), are consistent in finding that in the case of Mexico, two phenomena could be increasingly significant in the long term: an alteration in the cycle and volume of rainfall and an increase in the average temperature in some regions (Martínez 2010; Magaña 2007; Prieto 2007; Galindo 2009; IMTA 2010).

To the extent that the mean natural availability of water in Mexico is directly linked to weather cycles and that there is a pattern of contemporary land use that has led to greater population density and economic activities in the northern region, which is characterized by its low availability; considering the structure and organization of agri-food production in the country, the potential effect of a change in the rainfall cycle and potential drastic changes in volume would imply an impairment in the food production capacity, as can be seen in the results of the regression model (Table 12.1).

The possibility of Mexico to face chronic problems of water availability associated with climate change, defines a context for assessing the potential contributions of the water footprint approach. This approach can, in the process of

---

[1]Scenarios elaborated by the IPCC are a collection of models for the analysis of climatic implications under different combinations of economic growth, demographic dynamics, technological patterns, and specific energy sources. The scenarios are: A1–grouping the A1FI, A1T, A1B scenarios– , A2, B1, B2. These scenarios do not include conditions related to other climate initiatives such as those derived from the Framework Convention on Climate Change or the emissions standards associated with the Montreal Protocol (IPCC 2001).

strengthening institutional capacities, reduce the effects of an eventual decrease in the water volume for agricultural/livestock production. It should be noted that from the perspective of sectoral linkages observed in the Input-Product Matrix of Mexico and the ones derived from the application of the Rasmussen technique (SEMARNAT 2009), the food production sector in Mexico is a major transmission node of economic effects to other productive sectors.

12.3   Vulnerability and Water Footprint

As indicated in other chapters of this book, the methodological approach of water footprint and virtual water has a significant potential for the design of alternative institutional arrangements to water security based on the model of increasing extraction and overexploitation of water bodies above its natural balance.

It is more common to emphasize that water management has a transversal approach that requires management tools that are beyond the field of hydraulics strategy and environmental policy (Conagua 2012). Indeed, to the extent that water is an indispensable good of nature to sustain life, the articulation of productive activities, and maintenance of environmental services that directly affect social welfare, there are biophysical, social, economic, and institutional dimensions around it, which makes it an issue of the development agenda. It is required the coordination of the economic, hydraulic, environmental, economic promotion, health, urban planning, rural development, and social development policies. This set of issues on the formal government agenda influences the design of incentives and infrastructure financing, water resources management and safeguarding of its balance, its use patterns, effects on health of people and ecosystems, urban planning and correction of asymmetries in the public welfare, by reducing incentives to plunder the natural capital.

Until recently, water policy mainstreaming did not have tools to explore the processes of production linkages and articulation of public welfare that occurs from the way water is used. From this perspective, the potential associated with the calculation of water footprint and water colors allows a dimension of strengthening management capacities as they identify magnitudes of consumption and utilization associated with different availability sources (green water, blue water, and gray water). At the same time, it allows to direct efforts to promote technological change and required industrial organization processes to reduce the risks associated with water availability reduction.

Facing the presence of adverse social effects of the occurrence of natural events such as droughts and floods—besides those related to social patterns of water resource appropriation—the water footprint and virtual water approach is an instrument with a significant potential to be used to characterize the vulnerability that is not limited to geographical areas of impacts, but transcends through networks of inputs supply (Garrido 2010).

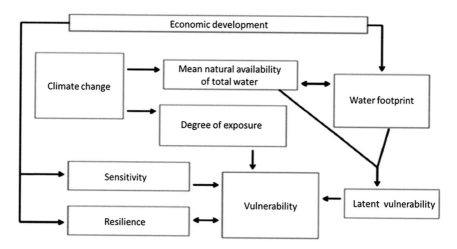

Fig. 12.1 Relationships between vulnerability, climate change, and water footprint. *Source* Own elaboration with data from Conagua

The vulnerability concept used here is convergent with the concept of water footprint impact used by Hoekstra (2008). For him, water footprint impact depends on many factors, including the water availability at local levels, competition among various uses of water, climate requirements, and assimilative capacity of the water system. Water footprint impact depends on the vulnerability of the region where a water use center is located.

The relationships between climate change, water footprint, and vulnerability are described in Fig. 12.1. Climate change affects water availability by altering the temperature, sea level, and precipitation in an area (Kundzewicz 2007); in the same way, it alters the degree of risk exposure by increasing the scale and presence of disasters such as droughts and floods. For its part, the action of man on nature, caused by socioeconomic development, has effects on the use of water resources and, therefore, on water footprint and vice versa. Meanwhile, human activity influences the degree of sensitivity to the impact of extreme events that is measured from socioeconomic assets or actors that are affected by disasters (Martínez 2010).

Water footprint and water availability allow to approach the calculation of the latent vulnerability[2] of a region, i.e., medium-term vulnerability. If an economy has a very high water footprint in relation to its average availability, an abrupt variation caused by a drought would impact more than the local resilience capacity to face it.

---

[2]The concept of latent vulnerability is used in the sense of underlying inflation in the economy. That is, a phenomenon that has not yet occurred but could be decisive in the amplification effects.

12.4 An Approach to Latent Vulnerability with Water Footprint Instruments

A methodology is presented to calculate vulnerability based on elements already known about water footprint.

If water footprint is defined as the water volume required (directly or indirectly) to produce goods and services consumed by the inhabitants of a certain geographical area, it may be originated by goods produced domestically or brought from abroad (Hoekstra and Chapagain 2007), as follows:

$$(1) WFP = IWFP + EWFP$$

where

IWFP   Internal water footprint, which is water coming from the national resources of a given geographical area.

EWFP   External water footprint, which is the water amount needed to produce goods or services consumed in a given geographical area, when these are produced abroad.

12.4.1 Internal Water Footprint (IWFP)

It is defined as the use of domestic water resources to produce goods and services consumed by the inhabitants of a certain geographical area; it is the sum of water amount used in different sectors.

$$(2) IWFP = AWU + IWW + DWW - VWE_{dom}$$

where

AWU   water used in agriculture (*Agricultural water use*)
IWW   water used in the industrial sector (*Industrial water withdrawal*)
DWW   water used in households (*Domestic water withdrawal*)
VWE_{dom} it is the virtual water export to other geographical areas (*Virtual water export related to export of domestically produced products*).

12.4.2 External Water Footprint (EWFP)

It is defined as the annual water resources volume used in other geographical areas to produce goods or provide services which are consumed in a given geographical area.

$$(3)\, EWFP = VWI - VWE_{re-export}$$

where

VWI   volume of virtual water imported from goods

$VWE_{re-export}$ volume of virtual water exported to other geographical areas as a result of the re-export of imported products.

12.4.3   Latent Vulnerability (LV)

It is defined as the ratio of water footprint volume and resource availability,

$$(4)\, LV = WFP/D$$

where

LV   Latent vulnerability
WFP   Water footprint
D   Mean natural availability of total water.

The higher the ratio, the greater is the vulnerability.

12.4.4   Internal Latent Vulnerability (ILV)

$$(5)\, ILV = (AWU + IWW + DWW - VWE_{dom/D})$$

where

AWU   it is the agricultural water uses (*Agricultural water use*)
IWW   it refers to industrial uses (*Industrial water withdrawal*)
DWW   it refers to domestic uses (*Domestic water withdrawal*)
VWE_{dom} it is the virtual water export to other geographical areas (*Virtual water export related to export of domestically produced products*).
D   it is the mean natural availability of total water.

12.4.5   External Latent Vulnerability (ELV)

$$(6)\, ELV = (VWI - VWE_{re-export/D})$$

where
VWI   volume of imported virtual water
$VWE_{re-export}$ volume of virtual water exported to other geographical areas as a
                    result of the re-export of imported products.
D   mean natural availability of total water.

Previous estimates can be made at different geographical scales, with respect to
different water types: green, blue, and gray; and for different sectors: agricultural,
industrial, and domestic.

One of the weaknesses of the water footprint concept identified by Brown and
Matlock (2011) is that the sum of water volumes used in different contexts cannot
be compared without standardization. The impact of the volume consumed in a
region with abundant resources is different from the same volume in a region with
scarce water resources.

An empirical approach to the latent vulnerability on water resources in Mexico is
presented. Federal states were taken as the analysis unit due to the nature of the
available information. The following variables were considered to quantify water
vulnerability:

1. Water volume allocated in the agricultural sector in cubic meters (water use in
   agriculture).
2. Total aggregated agricultural census gross value in thousands of pesos.
3. Average annual precipitation in the states, in cubic mm.
4. Irrigated surface in hectares.
5. Water volume assigned to irrigation surface in cubic meters.
6. Total rainfed surface in hectares.

Vulnerability assessment was conducted with statistical cluster analysis,[3] a
nonhierarchical method known as k-means, using the JMP program version 8.

Standardization process was used to avoid distortions that could be originated
using different scales and units of measurement; however, the results are presented
in unstandardized numbers to facilitate interpretation.

By taking as a reference the characteristics of the federal states regarding their
consumption of water allocated, average annual precipitation, harvested surface,
and rainfed surface, it was decided to have three classifying groups.[4]

---

[3]Although cluster analysis is purely descriptive, it is a useful statistical tool to define groups based
on similarities, it comprises techniques that produce classifications or types from data that were not
classified initially. The method implicates finding similarities between observations by measuring
the metric distance between them. There are two methods: hierarchical and nonhierarchical. With
the first one, data are grouped sequentially in a nested succession using the nearest neighbor
method. In the nonhierarchical method, a set of seed points is selected depending on the number of
selected clusters a priori and then build clusters around these points.

[4]In cluster analysis there is no single solution as the choice depends largely on the researcher
criterion and his/her theoretical framework. He/she faces a trade: as the average distance decreases,
the number of clusters increases and vice versa.

Table 12.2 Approach to latent vulnerability with water footprint instruments

Federal state	Cluster*	Average distance	Agricultural volume allocated (m³)	Agregated Census Gross Volume (thousands of pesos)	Average precipitation	Irrigated sown surface (ha)	Distributed volume (m³)	Total labor surface (ha)	Temporary total surface (ha)
Sinaloa	1	1.59	8607988788.0	1868633.0	711.1	820605.00	9158932000.00	1357581.60	536976.60
Sonora	1	1.59	6355568435.5	2537112.0	414.6	402385.00	3913682000.00	1510111.60	1107726.60
Chihuahua	2	2.42	4572200253.0	8299.0	485.8	93508.00	1589426000.00	1858789.90	1765281.90
Guanajuato	2	1.35	3436550479.0	19703.0	681.5	189461.00	1624574000.00	1045481.90	856020.90
Tamaulipas	2	2.24	3361881159.5	479320.0	784.9	359062.00	2516853000.00	1484165.80	1125103.80
Baja California	2	1.82	2514400000.0	528727.0	190	197364.00	2573712000.00	400396.20	203032.20
Hidalgo	2	0.65	2076041720.9	20731.0	969	88156.00	1604631000.00	597485.60	509329.60
Coahuila	2	0.99	1623737817.0	18993.0	373.7	3851.00	58162000.00	949095.50	945244.50
Colima	2	0.92	1579000000.0	188517.0	1020.00	28331.00	744176000.00	209521.80	181190.80
Nuevo Leon	2	0.52	1456554428.0	26958.0	575.5	11939.00	212913000.00	647061.30	635122.30
Durango	2	0.71	1382168417.0	4671.0	575.6	14538.00	169159000.00	984061.40	969523.40
Zacatecas	2	1.75	1323029092.0	2838.0	575	10295.00	1345920000.00	1766832.60	1756537.60
Mexico	2	0.46	1268594694.5	23298.0	913.6	22201.00	106471000.00	717386.10	695185.10
San Luis Potosi	2	0.66	1126380046.0	4529.0	900	73742.00	186194000.00	1061906.70	988164.70
Nayarit	2	1.12	1057738688.0	316460.0	1270.00	40184.00	483337000.00	618747.50	578563.50
Yucatan	2	0.63	960147688.0	261462.0	971.7	7592.00	49645000.00	628647.00	621055.00
Morelos	2	1	940594797.0	22657.0	966.7	21467.00	366761000.00	151272.60	129805.60
Queretaro	2	0.89	666851481.4	8538.0	669.4	8806.00	91755000.00	241075.80	232269.80
Campeche	2	1.21	582655212.0	311946.0	1240.90	0	0	843258.00	843258.00
Aguascalientes	2	1.11	486000000.0	8988.0	513.6	7591.00	73026000.00	174365.70	166774.70

(continued)

H.R. Dávila-Ibáñez and R.M. Constantino-Toto

Table 12.2 (continued)

Federal state	Cluster*	Average distance	Agricultural volume allocated (m³)	Agregated Census Gross Volume (thousands of pesos)	Average precipitation	Irrigated sown surface (ha)	Distributed volume (m³)	Total labor surface (ha)	Temporary total surface (ha)
Baja California Sur	2	1.92	3359000000.0	760807.0	190	28677.00	160813000.00	139071.50	110394.50
Tlaxcala	2	1.07	149773761.0	1613.0	813.3	6483.00	22871000.00	205614.60	199131.60
Quintana Roo	2	1.41	134843858.0	68312.0	1293.80	2600.00	16423000.0,	378091.90	375491.90
Michoacan	3	2.02	4620054919.0	241758.0	958.4	210034.00	2297825000.00	1477048.30	1267014.30
Jalisco	3	1.31	3171712983.5	152869.0	922.8	82406.00	704873000.00	1772292.20	1689886.20
Veracruz	3	1.8	2627221728.1	438207.0	1740.10	55845.00	848069000.00	2730129.90	2674284.90
Puebla	3	1.14	1955775129.0	14312.0	1221.70	21064.00	284339000.00	1022922.70	1001858.70
Chiapas	3	1.83	1406745938.5	262031.0	2214.40	37598.00	353797000.00	2252972.40	2215374.40
Oaxaca	3	0.82	868999613.6	151436.0	1217.50	22184.00	495133000.00	1681890.00	1659706.00
Guerrero	3	0.81	848472429.0	356538.0	1227.80	19923.00	352737000.00	1641717.30	1621794.30
Tabasco	3	2.37	174860039.1	469408.0	2211.80	0	0	622179.00	622179.00

Source

The following results can be observed:

- *Cluster 1*. High vulnerability, characterized by an average low precipitation (about 562 mm^3/year), high consumption of water allocated (around 7,480 cubic hectometers/year), which is destined to irrigated sown surfaces (an average irrigated surface of 611,495 hectares), besides having a high share of temporary agricultural surface (an average of 822,351.6). In this group only two states are found: Sinaloa and Sonora.
- *Cluster 2*. Medium vulnerability, which has the following characteristics: an average annual precipitation of 760 mm^3/year (equal to the national average), low consumption of water allocated (an average of 1,477 cubic hectometer/year), which is reflected in a reduced irrigated sown surface (an average of 57,897 hectares). However, they are states that exhibit medium vulnerability to climate change and, therefore, they are affected by a potential water shortage, as they have a considerable temporary agricultural surface (an average of 661,261 hectares). This group consists of 21 states.
- *Cluster 3*. Low vulnerability. In this group there are eight states where average annual precipitation is high (an average of 1,464 mm^3/year), with a low consumption of water allocated (an average of 1,959 cubic hectometers/year), with a low participation in irrigated surface in the country (an average of 561,131 hectares), but with a large temporary agricultural surface, with an average of 1.594 million hectares (Table 12.2).

The results of the vulnerability level of the federal states are consistent with the fact that the greatest resilience capacity to droughts occur where rainfall levels are abundant and the productive structure implies a low water allocation, with a low coefficient of participation in the irrigated surface of the country. However, the federal states that concentrate a significant proportion of national agricultural production exhibit a potential impact of the increasing drought, which would have to orient public efforts to reduce the incidence of underlying damage to the food production of the country.

References

Bitrán, Daniel. *Metodologías para la evaluación del impacto socioeconómico de los desastres Studies and Perspectives Series* 108. México: CEPAL 2009.

Brown, Amber and Marty D. Matlock. *A Review of water scarcity indices and metodologies. The sustainability Consortiun*, White Paper Num. 106, USA: University of Arkansas, 2011.

CONAGUA. *Hipercubo de estadísticas del agua en México*. México: SEMARNAT, 2011.

CONAGUA. *Agenda del agua 2030. Avances y logros 2012*. México: SEMARNAT, 2012.

Constantino, Roberto and Hilda R. Davila. "Una aproximación a la vulnerabilidad y la resiliencia ante eventos hidrometeorológicos extremos en México" in *Política y Cultura*, 38. México: Universidad Autonoma Metropolitana, 2011.

Cutter, Susan. et al. "A Place-based Model for Understanding Community Resilience to Natural Disasters" in *Global Environmental Change*, 18. USA: 2008.

Galindo, Luis Miguel. *Economía del cambio climático en México*. México: SHCP, SEMARNAT, 2009.

Garrido, Alberto et al. "Water footprint and virtual water trade in Spain" in *Natural Resource Management and Policy Series*. Spain: Springer, Fundacion Marcelino Botin, 2010.

Hoekstra, Arjen and Ashok K. Chapagain. "Water footprints of nations: Water use by people as function of their consumption pattern", *Water Resources Management* 21. (2007): 35–48.

Hoekstra, Arjen. *Water Neutral. Reducing and Offsetting the Impacts of Water Footprints*, Research Report 28. The Netherlands: UNESCO-IHE, 2008.

IPCC. *Climate change. Impacts, adaptation and vulnerability*, Third assessment report of the IPCC, WMO, UNEP: 2001.

Kundzewicz, Zbigniew et al. *Freshwater resources and their management. Climate Change 2007: Impacts, Adaptation and Vulnerability*, Contribution of Working Group II to the Fourth Assessment Report of the IPCC. Cambridge: University Press, 2007.

Magaña, Victor and Caetano, Ernesto. *Pronóstico climático estacional regionalizado para la República Mexicana como elemento para la reducción de riesgo, para la identificación de opciones de adaptación al cambio climático y para la alimentación del sistema: cambio climático por estado y por sector. Informe final del proyecto INE/A1-006/2007*. México: SEMARNAT, INE, UNAM, CCA, 2007.

Martínez, Polioptro. *Efectos del cambio climático en los recursos hídricos de México*. México: IMTA, 2007.

Martínez, Polioptro. *Altas de vulnerabilidad hídrica en México ante el cambio climático*. México: IMTA, 2010.

Moser, Sussane. *Resilience on the face of global environmental change*, CARRI Research Report 2. USA: Oak Ridge National Laboratory, 2008.

Perrings, Charles. "Resilience and sustainable development", *Environment and Development Economics* 11. UK: Cambridge (2006): 417–427.

Prieto, Ricaro et al. *Determinación de periodos de sequía y lluvia intensa en diferentes regiones de México ante escenarios de cambio climático*, Informe final del proyecto: INE/A1-056/2007. México: SEMARNAT, INE, IMTA, 2007.

SEMARNAT. *Distribution of costs of climate change among sectors of the Mexican economy. An input-output approach, Mexico*. México: SEMARNAT, 2009.

Vargas, Jorge E. "Políticas públicas para la reducción de la vulnerabilidad frente a los desastres naturales y socionaturales", *Environment and Development Series* 50. Chile: CEPAL, 2002.

Part III
Methodology for Analyzing Water Footprint and Virtual Water

Chapter 13
Water Demand of Major Crops: A Methodology

Ignacio Sánchez-Cohen, Ernesto Catalán-Valencia
and Jaime Garatuza-Payán

Abstract An algorithm to compute crop water requirements in the irrigation districts of Mexico is presented based on climatological data from where irrigation needs are obtained. The method relies on the soil water balance where the main inputs are rainfall and irrigation and the outputs are deep drainage or deep percolation, and crop evapotranspiration. The method to compute each variable of the soil water balance is outlined as well as irrigation needs. A computer program was utilized, DRIEGO1, from where crop's water needs are deduced. The drought index and evapotranspiration data for the main irrigation districts of the country are presented in both graphical and tabular ways. According to the results it can be seen that the crop evapotranspiration is strongly dependent on the climatology showing significant difference for the same crop for different places.

Keywords Evapotranspiration · Irrigation water use

13.1 Introduction

The real objective of irrigation is to obtain the highest possible performance. For watering to be profitable, the increase in profits for its effect should be greater than the total annual cost of irrigation; this is especially important in ecosystems where the main water source is the aquifer and pumping is required. In these cases, it is more attractive to irrigate high-value crops, such as fruit and vegetables, than annual crops; however, it is necessary to study the production chain, from cultivation all the way to retail, to define the real expectation of technified irrigation. It is recognized that the 'best' irrigation system does not guarantee profits if it does not have appropriate fieldwork for soil management, seed quality, fertilization, etc. ("Good management practices") for the crop to grow without problems.

The producer interested in technified irrigation should consider the following questions: What is the effect of irrigation on production? How much water is needed? Does the water source can guarantee this quantity? Is the water quality satisfactory?

When watering, water infiltrates gradually through the crop root zone. For water to move across soil layers, the first layer must be "filled" to field capacity, i.e., once this layer retains as much moisture as possible according to its characteristics. This water movement is known as "wetting front" advance. Water moves much faster in light-textured soil, such as sand, than in one with heavy or clayish texture.

When sufficient irrigation water was supplied to wet the root zone, the plant transports it through the stems to the leaves and fruits. The leaves have thousands of microscopic openings called "stomata," through which the plant loses water vapor. This continuing loss is known as "transpiration" and can cause plant death if irrigation is not applied in a timely manner and in the right quantities.

Water requirement for plants is the amount of water lost by transpiration plus the amount evaporated directly from the soil; these two processes are known as "evapotranspiration." Evapotranspiration rates vary and depend on day length, temperature, cloud cover, wind speed, relative humidity and type, size, and number of plants per hectare.

Water is essential for the development of physiological processes of all plants. It is the primary medium for chemical reactions and movement of substances across the various parts of the plant. Also, it is a vital element in photosynthesis and metabolism, including cell division and growth of cells and it is the means by which plants are kept fresh through the process of transpiration.

Water is the primary factor that determines the crop yield. If a crop without moisture closes its stomata, it will roll its leaves reducing the growth of its parts and affecting greatly its performance. So, the purpose of irrigation is to provide timely adequate amount of water to crops to prevent damage that impact performance. Therefore, producers must obtain answers to the questions how, how much, and when to water. This chapter emphasizes the last two questions.

13.2 Agro-ecological Zones and Water Availability

All crops require different amounts of water; the amount they need during their growth period is called "seasonal evapotranspiration" or "seasonal consumptive use." This varies with the climate of the different areas of the country; in arid zones, crops require more water than in humid zone, mainly because precipitation rates are low and evaporation rates are high. Figure 13.1 shows the relationship between

Fig. 13.1 Graphical
representation of usable soil
moisture. *Source* The authors

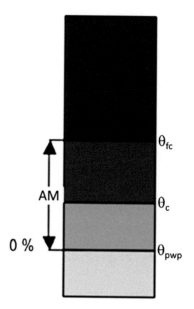

precipitation and evaporation in the irrigation districts of the country. This relation
is an indicator of the different water availability and it serves, indirectly, to infer
irrigation needs.

13.3   Methodology to Meet Crop Water Needs

In this section, the methodology described in Catalán (2002) and Catalán et al.
(2005) is used. Irrigation scheduling is a process oriented to determine the amount
of water to be used and the dates of each irrigation to minimize deficiencies or
excess soil moisture that may affect the growth, yield, and quality of crops. Proper
irrigation scheduling can reduce costs by saving energy and labor force, minimize
water stress, and maximize yields, quality, and profitability of the crop.

The most used methods for irrigation scheduling are of two types: (*a*) those that
measure and monitor some variables related to crop water stress in soil or plants and
(*b*) those that estimate the moisture balance in the soil. The former rely on the use of
sensors to measure variables such as moisture content, moisture tension, and
electrical resistivity of soil or foliage temperature (Martín et al. 1990). The main
calculations of the present computational application are based on method
(*b*) which estimates the components of the water balance in the soil.

13.4 Moisture Balance in the Soil

This method consists in making an assessment of the amount of water present in the soil profile occupied by the crop roots in a period of time. According to the principle of conservation of mass, the amount of water that enters minus the amount that exits in a period equals the change in water storage experienced by the soil profile during that period

$$\Delta\theta = R + P - ETr - D \qquad (13.1)$$

where

$\Delta\theta$ is the change of water content in the soil during the period considered,
R is the amount of irrigation water,
P is the precipitation or rainfall,
ETr is evapotranspiration or water consumption, and
D refers to deep drainage

Each term in the equation must be quantified in the adequate timescale and space. The spatial scale is the volume control limited by the depth of soil explored by the crop roots, and the time scale is variable, being the daily scale the one used more often (Fox et al. 1994; Ojeda et al. 1999; Catalán 2002).

13.4.1 Soil Moisture Content

Irrigation water used in agriculture infiltrates into the soil saturating the surface layer. Then, it is transmitted and redistributed to greater depths until a portion of the water drains from the soil profile occupied by the roots. The speed of this movement depends on soil hydraulic properties, which are related to other properties such as texture, being higher in light sandy soils than in heavy clayey soils. As a result of this movement and water consumption by evapotranspiration, the moisture content in the soil decreases with time.

13.4.2 Field Capacity

The application of irrigation is not intended to saturate the soil profile, but to raise its moisture content to an optimum level known as field capacity (θ_{fc}). It is defined as the maximum amount of water that the soil can hold against gravity, after being saturated and in absence of direct evaporation. This condition is achieved in a period of 3–10 days depending on the soil type and its ability to retain water. In practical terms, θ_{fc} is the moisture content that is achieved in the soil after the downward movement or drainage of water has decreased to a level which could be considered a low or negligible loss of water.

13.4.3 Permanent Wilting Point

Irrigation must be applied before the available soil moisture is depleted completely, taking as a reference a minimum allowable moisture content for plants: the *permanent wilting point* (θ_{pwp}). Below this moisture content, certain plants or crops no longer recover their turgidity, even after being in a saturated atmosphere for 12 h.

13.4.3.1 Available Moisture

The moisture range between θ_{fc} and θ_{pwp} is known as the available moisture (AM) for plants and it refers to the maximum amount of water that can hold the soil profile occupied by the roots of plants (Pr).

Regularly, it is estimated as a sheet or thickness of water:

$$AM = (\theta_{\mathrm{fc}} - \theta_{\mathrm{pwp}})Pr \qquad (13.2)$$

θ_{fc} and θ_{pwp} are expressed in cubic meters (m^3) of water per m^3 of soil; *AM* and *Pr* in meters. Table 13.1 shows average values of moisture content at saturation (θ_{s}), θ_{fc}, θ_{pwp}, and AM for each type of texture and a meter of soil depth, which were obtained from Saxton et al. (1986).

Table 13.1 Moisture constants by soil type

Texture	θ_{s}	θ_{fc}	θ_{pwp}	AM
Sand	0.3545	0.128	0.0567	0.0714
Loamy sand	0.3878	0.1598	0.0764	0.0834
Sandy loam	0.4697	0.2522	0.174	0.0782
Loam	0.4617	0.254	0.118	0.136
Sandy clay loam	0.4784	0.2676	0.1724	0.0952
Clay loam	0.5018	0.3215	0.1838	0.1377
Silty clay loam	0.5203	0.3648	0.1941	0.1707
Silt loam	0.4676	0.2857	0.1062	0.1794
Sandy clay	0.5052	0.3333	0.2419	0.0914
Silty clay	0.5422	0.4403	0.2786	0.1617
Clay	0.5566	0.5359	0.4127	0.1232
Silt	0.4154	0.3154	0.0962	0.2192

Source The authors

13.4.4 Abatement of Available Moisture as an Irrigation Criterion

Irrigation is applied when the soil moisture content decreases to a critical value (θ_c) that determines the maximum degree of water stress to which the crop is subjected. Usually, this value is estimated in the range of the available moisture, where θ_{pwp} and θ_{fc} represent zero and 100 % of AM, respectively (Fig. 13.1). A fraction or percentage of maximum abatement of available moisture, FAM, is used

$$\theta_c = \theta_{fc} - \frac{FAM}{100}\left(\theta_{fc} - \theta_{pwp}\right) \tag{13.3}$$

where FAM is expressed in percent; the remaining terms have been defined previously.

Soil water content on a particular day, θ_i, is estimated based on the water content of the previous day, θ_{i-1}, and the remaining terms of Eq. (13.1) estimated for the current day:

$$\theta_i = \theta_{i-1} + R_i + P_i - ETr_i - D_i \tag{13.4}$$

13.4.5 Water Consumption by Evapotranspiration

Irrigation water is consumed by the evaporation that occurs from the top soil and by transpiration from the surface of the leaves and that has been absorbed previously by the roots of the plant. Given the difficulty of measuring separately evaporation and transpiration, the term evapotranspiration (ETr) is used to refer to both types of water flow (FAO 1989).

The amount of water consumed by a crop during its cycle is directly related to its performance (Doorenbos/Kassam 1996). The value of ETr at a given time depends on climatic factors such as temperature and air humidity, solar radiation, and wind speed. It also depends on the phase or degree of development of the crop and some physiological mechanisms that control plant responses to changes in environmental conditions (Catalán et al. 2004).

13.4.6 Estimation of Evapotranspiration

There are several methods to estimate ETr both directly and indirectly. Direct measurement is based on the use of instruments such as the lysimeters to monitor changes in soil moisture content over time. Indirect methods use physical-empirical formulas that require data from climate and crops, and that vary in precision depending on the approximation approach and the type and number of variables involved (Sánchez et al. 2006). In this case, an indirect method was used to calculate ETr based on the previous estimation of the reference evapotranspiration (ET0)

$$ETr = K_s K_c ET_0 \qquad (13.5)$$

where ET_0 is the potential ET of a reference crop, alfalfa or grass, well irrigated without water limitation (Allen et al. 1990; Jensen et al. 1990). ET_0 is estimated with the original method of Hargreaves (1974), later modified by Hargreaves/Samani (1982), which requires the latitude and daily air temperature (minimum and maximum). K_c is an empirical dimensionless coefficient for a specific crop, a given state of growth, and a particular condition of soil moisture; it indicates the relative capacity of the soil surface and crop to match the evaporative demand of the soil surface and reference crop under the same climatic conditions (Jensen et al. 1990). Its values are obtained from a specific K_c curve for each crop, which describes the variation of this coefficient from the crop growing season (Doorenbos/Pruitt 1977). The $K_c ET_0$ product is also known as maximum ET of the crop, restricted only by environmental conditions. K_s is a dimensionless factor that restricts the maximum water consumption of the crop because of reduction or abatement of soil moisture and resistance to water flow that it causes. K_c reflects irrigation management, since the greater the spacing between waterings the higher its restrictive effect on water consumption.

13.4.7 Estimation of Effective Rainfall and Water Losses by Percolation

The term precipitation (P) of Eq. (13.1) is estimated as effective rainfall, since part of the rainfall is lost by evaporation or run-off and it is not available to plants. Due to the complexity of these processes, effective rainfall is estimated by empirical functions obtained from statistical analysis. Water losses by drainage or deep percolation occur when the capacity storage of water in the soil profile is exceeded, i.e., when P-ETr is greater than AM.

13.4.8 Computer Program

The Driego computer program (Sánchez/Catalán 2006), which systematizes the algorithm described, was used to estimate water requirements of crops. The figures below show water requirements of the major crops in the most important irrigation districts of the country in three major regions: arid subtropics, semiarid subtropics, and humid subtropics (Figs. 13.2, 13.3 and 13.4).

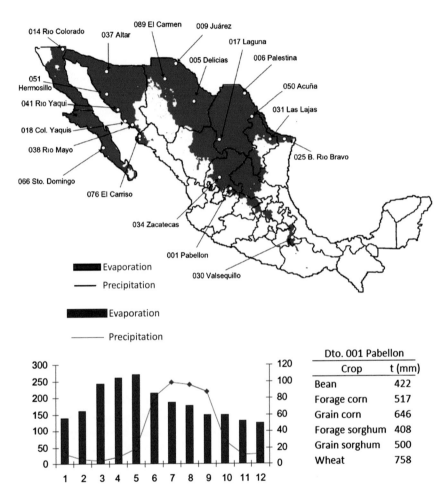

Fig. 13.2 Precipitation/evaporation and crop water demands balance in irrigation districts in the arid subtropics. *Sources* The Authors information

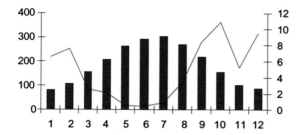

Dto. 014 Rio Colorado	
Crop	Et (mm)
Alfalfa	1994
Cotton	690
Safflower	623
Forage sorghum	435
Grain sorghum	504
Wheat	679

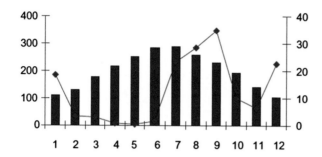

Dt. 066 Sto. Domingo	
Crop	Et (mm)
Alfalfa	2193
Chili pepper	707
Bean	549
Chickpea	494
Corn	976
Orange tree	1629
Tomato	860
Wheat	764

Dt. 017 Laguna	
Crop	Et (mm)
Alfalfa	1964
Cotton	964
Chili pepper	607
Bean	468
Forage corn	609
Grain corn	758
Walnut	1197
Forage sorghum	492
Grain sorghum	548
Wheat	717
Grapevine	1128

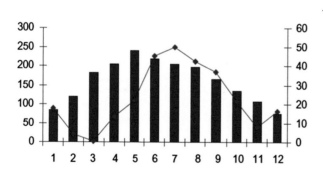

Fig. 13.2 (continued)

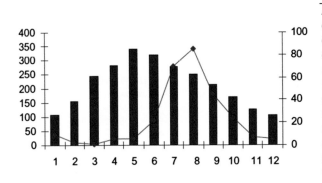

Dt. 005 Delicias	
Crop	Et (mm)
Alfalfa	1843
Cotton	689
Oat	560
Chili	847
Bean	444
Forage corn	664
Grain corn	909
Walnut	1176
Forage sorghum	530
Grain sorghum	474
Soy	417
Wheat	524

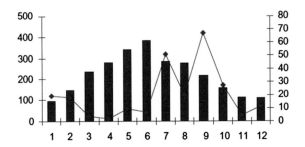

Dt. 009 Cd. Juarez	
Crop	Et (mm)
Alfalfa	1666
Cotton	632
Oat	593
Forage sorghum	398
Grain sorghum	398
Wheat	673

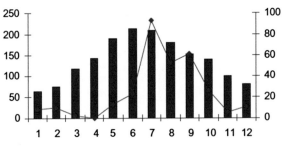

Dt. 089 El Carmen	
Crop	Et (mm)
Alfalfa	1661
Chili pepper	641
Bean	458
Grain corn	708
Grain sorghum	510

Dt. 031 Las Lajas	
Crop	Et (mm)
Bean	456
Grain corn	745
Forage sorghum	396
Grain sorghum	453
Wheat	620

Fig. 13.2 (continued)

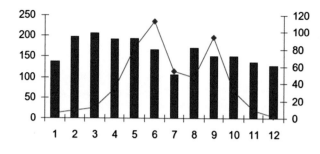

Dt. 030 Valsequillo	
Crop	Et (mm)
Chili pepper	500
Bean	462
Grain corn	977
Forage corn	799
Green tomato	249
Red tomato	399

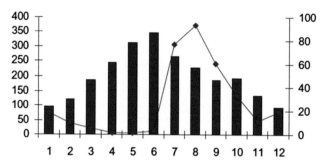

Dt. 076 El Carrizo	
Crop	Et (mm)
Sorghum	626
Soy	419

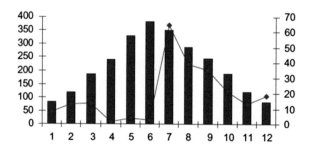

Dt. 037 Altar	
Crop	Et (mm)
Grain corn	676
Forage sorghum	478
Wheat	712
Grapevine	1369

Dt. 038 Rio Mayo	
Crop	Et (mm)
Cotton	613
Safflower	449
Chili	582
Bean	434
Chickpea	331
Grain corn	649
Grain sorghum	453
Soy	326
Wheat	750

Fig. 13.2 (continued)

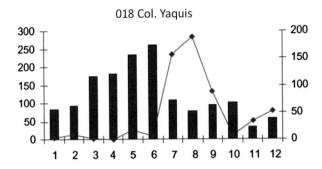

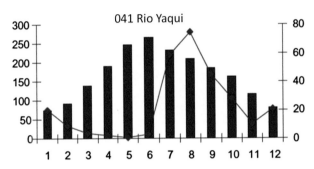

Dt. 018 Col. Yaquis 041 Rio Yaqui	
Crop	Et (mm)
Safflower	477
Chili	571
Chickpea	361
Grain corn	829
Soy	355
Wheat	809

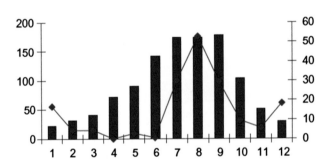

Dt. 051 Hermosillo	
Crop	Et (mm)
Bean	460
Chickpea	397
Grain corn	769
Wheat	797
Grapevine	1240

Dt. 025 Rio Bravo	
Crop	Et (mm)
Cotton	524
Bean	339
Grain corn	765
Grain sorghum	430
Wheat	534

Fig. 13.2 (continued)

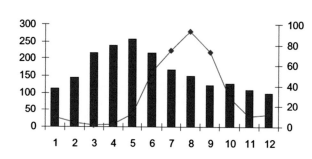

Dt. 034 Zacatecas	
Crop	Et (mm)
Alfalfa	1888
Oat	712
Chili pepper	813
Bean	494
Forage corn	883
Grain corn	707
Forage sorghum	592
Wheat	753
Grapevine	929

Fig. 13.2 (continued)

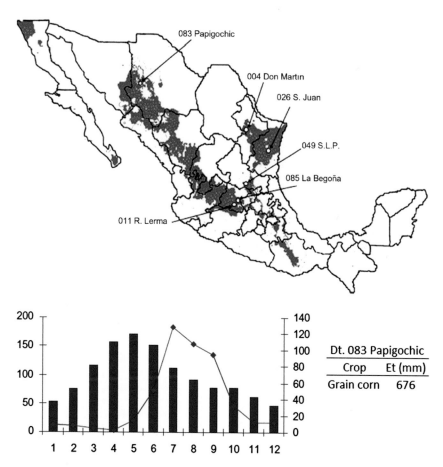

Dt. 083 Papigochic	
Crop	Et (mm)
Grain corn	676

Fig. 13.3 Precipitation/evaporation and crop water demands balance in irrigation districts in the semiarid subtropics. *Sources* Authors information

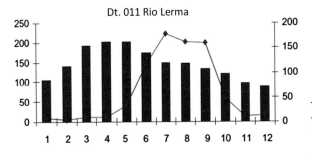

Dt. 025 Rio Lerma 085 La Begona	
Crop	Et (mm)
Chili pepper	580
Bean	479
Corn	639
Forage sorghum	362
Grain sorghum	431
Wheat	559

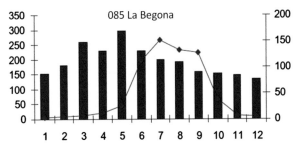

Dt. 004 Don Martin	
Crop	Et (mm)
Oat	560
Safflower	412
Bean	466
Grain corn	741
Forage sorghum	407
Grain sorghum	465
Wheat	640

Dt. 017 Laguna	
Crop	Et (mm)
Alfalfa	2049
Cotton	569
Rice	688
Oat	706
Safflower	519
Chili pepper	507
Bean	520
Chickpea	532
Corn	874
Forage sorghum	496
Grain sorghum	565
Soy	422
Wheat	777

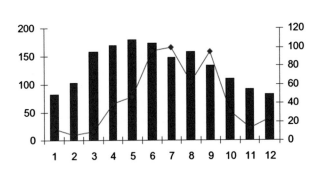

Fig. 13.3 (continued)

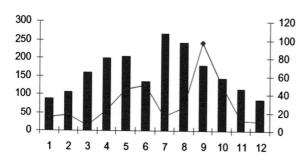

Dt. 026 San Juan	
Crop	Et (mm)
Cotton	684
Jalapeno pepper	572
Serrano pepper	605
Bean	428
Grain corn	753
Forage sorghum	370
Grain sorghum	567
Soy	427

Fig. 13.3 (continued)

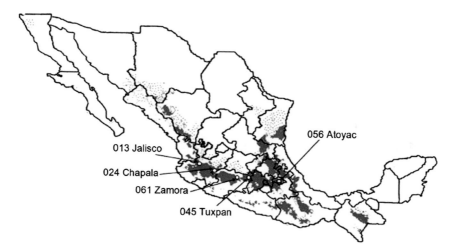

Fig. 13.4 Precipitation/evaporation and crop water demands balance in irrigation districts in the subhumid subtropics. *Sources* Authors information

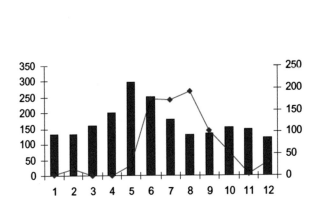

Dt. 013 Jalisco	
Crop	Et (mm)
Alfalfa	2076
Rice	544
Oat	550
Pumpkin	125
Safflower	659
Onion	922
Chili	236
Bean	462
Chickpea	461
Tomate	317
Lime	1516
Forage corn	673
Grain corn	1039
Forage grass	1722
Watermelon	231
Forage sorghum	307
Grain sorghum	646
Tomato peel	245
Wheat	883

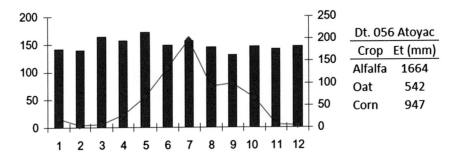

Dt. 056 Atoyac	
Crop	Et (mm)
Alfalfa	1664
Oat	542
Corn	947

Fig. 13.4 (continued)

13.5 Conclusions

Knowledge of crop water needs is an indicator of the amount of irrigation that should be applied to avoid undermining agricultural production. In terms of water footprint, this value is essential for its calculation because it quantifies the amount of water needed to produce the food on which our lives depend. The amounts of water consumed by crops depend, as already seen, on the locality and crop management, and they are connected to a chain of impacts on water resources in the production sites such as depletion in availability and water quality (Chapagain et al.

2006). In turn, the decline in availability is related to water productivity linked to irrigation efficiencies; thus, efficient irrigation can greatly reduce the amount of water used for a given crop.

The algorithm described here can be used as a tool for planning and decision making on the use of water resources. For example, to optimize crop pattern in the irrigation districts or units, i.e., to define the areas for each crop to produce the highest possible return, considering the volume of water available and other economic, ecological, and social restrictions.

References

Catalán A. Ernesto. *Programa para la calendarización del riego parcelario*, Informe de Investigación. México: CENID RASPA INIFAP, 2002.

Catalán A. Ernesto et al. "Aplicación computacional en red para la estimación de las demandas de agua y la calendarización de los riegos de los cultivos en los distritos de riego del país" *AGROFAZ* 5 (2005): 51–58.

Catalán, Ernesto, Vincent Gutschick and Magdalena Villa. "Análisis del control fisiológico sobre la transpiración y asimilación de especies forestales", Paper presented at the VII Congreso Internacional en Ciencias Agrícolas, Mexicali, Baja California. Febrero 25–28, 2004.

Chapagain, Ashok et al. "The water footprint of cotton consumption: An assessment of the impact of worldwide consumption of cotton products on the water resources in the cotton producing countries" *Ecological Economics* 60: 1 (2006): 186–203.

Doorenbos, Jan and Willian O. Pruitt. *Crop water requirements. Irrigation and Drainage* Paper 24. Rome: FAO, 1977.

Doorenbos, Jan and Amir Kassam. *Yield response to water.* Rome: FAO Irrigation and Drainage Paper 33, 1996.

FAO. *Irrigation water management: Irrigation scheduling*, Training Manual 4. Rome: FAO, 1989.

Fox, Fred et al. *Arizona Irrigation Scheduling* (AZSCHED Version 1.1E), Users Manual. USA: Cooperative Extension, University of Arizona, Tucson, 2004.

Hargreaves, H. George and Zohrab A. Samani. "Estimating potential evapotranspiration", *Journal of Irrigation and Drainage.* Div., ASCE, 108 (1982): 223–230.

Jensen, E. Marvin, Burman Robert and Richard, Allen. *Evapotranspiration and irrigation water requirements*, American Society of Civil Engineers, Irrigation Water Requirements Committee of the Irrigation and Drainage Division. Manuals and Reports on Engineering Practice No. 70. New York, 1990.

Martín, L., Stegman Earl and Elías, Fereres. "Irrigation scheduling principles", in *Management of farm irrigation systems*, edited by Hoffman, G. J., and K. H. Solomon ASAE monograph, 155–203, St. Joseph, MI., 1990.

Ojeda B., Waldo et al. *Pronóstico del riego en tiempo real.* México: IMTA, Centro Nacional de Transferencia de Tecnología de Riego y Drenaje, 1999.

Sánchez, Ignacio, Ernesto Catalán and Magdalena, Villa. "Evapotranspiration modeling for irrigation purposes. Chapter book", in *Modeling and Remote Sensing Applied to Agriculture,* 71–89, United States/Mexico: USDA ARS/INIFAP, 2006.

Saxton, Keith et al. "Estimating generalized soil-water characteristics from texture" *Soil Science Society of America Journal* 50 (1986): 1031–1036.

Chapter 14
Gray Water Footprint and Water Pollution

Anne M. Hansen

Abstract This chapter starts with the description of water pollution in relation to its use, the interrelation of contaminants with the water environment and the existence, as well with the lack of pollutant concentration regulations in water and its environment. It is described how the study of pollutant dynamics and decision-making for its control in the water environment can be performed with greater success when considering the entire basin. The gray water footprint concept is summarized, defined by Hoekstra/Chapagain (2008), and then some situations are discussed where, from the point of view of the author, this tool may use to better understand the impact of direct and indirect pollution caused by men and to obtain useful values for decision-making regarding pollutant loads and uses of water in basins.

Keywords Hydrologic basins · Water pollution assessment · Water pollution regulation · Grey water footprint · GWF application areas

14.1 Introduction

Pollutant accumulation into water bodies may represent a risk to biota and water as a source of supply for various uses. Many pollutants have high affinity to the mineral components or organic matter in sediment and soil so that the interaction of these solids with contaminated water can cause a cumulative process of contaminants. Other processes, that alter water features, can reverse this balance, causing the release of contaminants accumulated in sediment or soil. Thus, sediment and soil act as contaminant reservoirs, from where they can be released back into the water and become available, accumulate in the biota, or be eliminated through biological or photochemical degradation processes, form degradation products or metabolites, or be completely mineralized.

Water quality requirements vary according to the use it is given, including aquatic life protection, recreation, and water as a supply source for different uses. Sediment quality is almost always related to ecological health. Although the interactions

© The Author(s) 2016
R.H. Pérez-Espejo et al. (eds.), *Water, Food and Welfare,*
SpringerBriefs in Environment, Security, Development and Peace 23,
DOI 10.1007/978-3-319-28824-6_14

154

154   A.M. Hansen

between water and sediment are dynamic processes, which vary according to the different conditions of the water body, there are currently no regulations that relate quality of sediment and water.

14.2 Hydrologic Basins

The "Hydrologic Basin" is the basic unit for water management in Mexico and, therefore, for the assessment of water. It is defined as the surface that receives water, which flows to a collector, whether a canal or a river. "Hydrographic basins" are separated from each other by water catchment divisions. When the water comes from underground streams that are independent of surface structure, hydrologic basins are distinguished from topographic ones. This is particularly the case in limestone regions, where the hydrologic basin area often exceeds that of the hydrographic basin (Mauch et al. 2000).

When contaminants are introduced into a river, they are transported and transformed by physical, chemical, biochemical, and biological processes. Therefore, to determine the impact of these substances on water, the processes of transport, their transformation, and rates at which they are removed must be known (Chapman 1996).

Strategies for assessing water quality in Mexico are related to the different uses it is given, considering both ecosystems and society. It must be considered that there may be multiple uses of water in a hydrologic basin and that each use involves different requirements for its quality, which can lead to conflicts between different users. Ideally, water quality must meet the most demanding requirements for use as is the provision of good quality water for its purification and healthy ecosystems. It should be the responsibility of upstream users to ensure adequate water quality for downstream users in the same hydrologic basin.

14.3 Water Pollution

Water does not exist in pure form in nature given its ease to dissolve chemical substances. However, the main causes of water pollution are various human activities. Water pollution usually refers to the presence of contaminants coming from anthropogenic sources, causing it to be unfit for human consumption or to sustain aquatic life; but some natural phenomena such as volcanic activity, storms, or earthquakes can also cause changes in water quality.

Water pollution occurs when pollutants are discharged from individual and diffuse sources to water bodies, causing physical, chemical, and biological changes that produce adverse effects on humans and ecosystems. Individual sources of pollution are those that enter the water body by a pipe or channel discharge; they include municipal and industrial wastewater discharges, with or without pretreatment, or urban runoff drains. Diffuse sources refer to pollution that does not come from a single source; they are the accumulation of pollutants from runoff of areas with different land uses and can cause water eutrophication, which refers to the increase in the concentration of nutrients, which in turn can increase the primary productivity in water bodies, causing anoxic conditions and decreasing water quality, affecting the ecosystem and other water uses.

Despite the importance of water and sanitation, worldwide there is a population of 2.6 billion people without access to basic sanitation while 884 million have no access to drinking water (UNICEF/WHO 2010). The substances that contaminate water are organic and inorganic chemicals as pathogens. Some are found naturally in water bodies and their concentration is key in determining their natural origin or classification as a pollutant. Many chemicals are toxic and some of them are degradable, thus consuming the oxygen dissolved in water. These substances can be man-made or have natural origin, such as plant residues.

Pathogens can cause diseases in humans or animals. Each year, 1.4 million children die from diarrheal diseases, 88 % of them related to contaminated water, inadequate sanitation, or insufficient hygiene (UNICEF/WHO 2009). In addition to this, each year 860,000 children die from malnutrition, 50 % of them due to diarrhea or infections, also caused by contaminated water, inadequate sanitation, or insufficient hygiene (De Onis et al. 2004). Children are more vulnerable than other population groups to diseases caused by exposure to chemicals, which in total cause 4.9 million deaths annually, some of them from exposure to contaminated water (Prüss-Ustün et al. 2011).

14.4 Regulation and Assessment of Pollution in Water Bodies

Water bodies are exposed to thousands of pollutants from industrial, pharmaceutical, agricultural, and natural sources. Therefore, countries have implemented programs to regulate water quality to prevent contamination by substances with unacceptable concentrations.

Water politics in our country establish the following priorities: 1. To have sufficient water with appropriate quality, 2. To recognize the strategic value of water, 3. To use water efficiently, 4. To protect water bodies, and 5. To ensure sustainable development and environmental conservation (Conagua 2008).

The National Water Law (DOF 2012) states that water quality requirements depend on its use and that human consumption has priority over other uses. The standards NOM-127-SSA1-1994 (Permissible limits of quality and treatments to which water should be subjected to purification) and NOM-179-SSA1-1998 (Monitoring and evaluation of quality control of water for human use and consumption, distributed by public supply systems) set limits for human use and consumption. For their part, the NOM-001-SEMARNAT-1996 sets limits for discharges to waters and domestic goods, and NOM-002-SEMARNAT-1996 sets limits for discharges to urban and municipal sewage systems. The ecological criteria for water quality, CE-CCA-001/89, include limits for urban public use, recreation with direct contact, irrigation, livestock, and aquatic life.

Hansen/Corzo (2011) highlight the priorities and needs for the evaluation of pollution in hydrologic basins, referring to the water management policy, with its laws and regulations, in Mexico. They mention that the National Programme for Monitoring and Assessing (Proname in Spanish) persistent and bioaccumulative toxic substances (PBTS) is just being implemented; hence, there is no inventory or even formal assessments of exposures and risks associated with these substances. They propose a methodology for selecting substances in a PBTS monitoring program for water (Oswald-Spring 2011).

14.5 Gray Water Footprint Definition

The term gray water footprint (GWF) was introduced by Hoekstra/Chapagain (2008) to indicate the degree of water pollution. It is defined as the water volume required to assimilate a contaminant to acceptable water quality standards as defined in the criteria and standards.

Unlike "virtual water," GWF is a multidimensional indicator, which not only refers to the volume of contaminated water but also the place where it occurs (Hoekstra et al. 2011). It indicates the degree of water pollution in a particular process, site, or hydrologic basin. Although it is defined as the water volume required to assimilate the pollution load, it is calculated as the water volume required to dilute pollutants until concentrations are below the limits established in water quality standards.

GWF is calculated by dividing the amount of pollutant (L, mass/time) by the difference between the maximum allowable concentration for this pollutant (Cmax mass/volume) and its natural concentration in the receiving water body (Cnat, mass/volume) (Hoekstra et al. 2011):

$$GWF = L/(Cmax-Cnat)$$

The natural concentration of dissolved substances in water systems is the concentration that it would have in the absence of human intervention. Assimilative capacity of water with respect to a pollutant depends on the difference between the desired concentration (e.g., the maximum allowable concentration) and the contaminant natural concentration.

GWF can be determined directly from pollutants that are discharged into surface water bodies. However, when contaminants are discharged to the ground, as in the case of fertilizers or pesticides, a fraction is attenuated in the soil, while others drain superficially or infiltrate the soil into groundwater. In these cases, the pollution load and the corresponding GWF is the sum of the infiltrated and drain fractions.

14.6   Application Areas

As with blue and green water footprints, GWF can be calculated for a product, a person, or an area. GWF of a product indicates the water volume that pollutes the entire production chain. Thus, the specific processes where GWF can be reduced can be identified. This allows to inform consumers about the water volume contaminated during product manufacture and serves to know how much contaminated water is bought or sold when importing and exporting products.

On the other hand, GWF of a consumer, a consumer group or a company, serves to illustrate which process pollutes more water. Usually food has the greatest impact on water footprint (Dourte/Fraisse 2012).

An area can represent a municipality, a hydrologic basin or a country. GWF allows to estimate the contaminated water inside and outside the area. This allows to calculate the purchase and sale of contaminated water from product trading. GWF of culture allows to compare agricultural systems in different regions, different agricultural practices and management systems.

Currently, the concept of GWF has been used to calculate the water footprint resulting from contamination by total nitrogen and phosphorus of the major hydrologic basins in the world (Liu et al. 2012). These authors found high levels of pollution in rivers of tropical and subtropical areas and they also found that the main problems for nitrogen and phosphorus are located in the Southern Hemisphere. The results suggest that the degree of contamination of rivers with nutrients is increasing.

Next, there are some thoughts for other applications of this tool.

14.6.1   GWF of Reactive Contaminants

Currently, there is a lack of methods to describe the destiny of other pollutants in water systems, such as reactive transport processes that include transformations by physical, chemical, biological, and biochemical processes or "natural attenuation" of these substances.

"Natural attenuation" of contaminants refers to their concentration decrease by processes such as advection, dispersion, dilution, diffusion, volatilization, ion exchange, formation of complexes, abiotic transformation, and biological processes such as biodegradation and incorporation in the food chain (USGS 2012). Most contaminants tend to be adsorbed in solids suspended in water or in sediment particles, causing resistance to biodegradation, reducing volatilization, and affecting accumulation in aquatic organisms. The solubility of contaminants determines the degree of partition between sediment, interstitial water and water column. As water is a polar solvent, polar and ionic compounds dissolve more in it than nonpolar contaminants, which have a greater affinity to organic matter suspended in the water column or deposited in the sediment. Biodegradation rate indicates how fast the substance is decomposed by biological processes. Although the bioaccumulation factor is not a chemical property, it helps to estimate the degree of incorporation of contaminants to the tissue of aquatic organisms. These processes influence the attenuation of contaminants and can be used to predict their transport, environmental fate and the GWF they cause.

Transport mechanisms of contaminants adsorbed in sediment or incorporated into organic matter depend on the type of water body, flow rate, and sediment characteristics. Contaminants associated with sediment and organic matter are transported by suspension or bottom sediment transport. Clay and silt-size particles are generally transported by suspension in the water column as suspended solids. Dissolved contaminants are transported with the water, while immiscible liquid contaminants float or sink, depending on their density. When the latter have lower densities than water, they tend to float and become susceptible to volatilization or photolysis. When immiscible contaminants are denser than water, they are transported near the bottom and can be incorporated into the sediment phase.

Biological processes also affect the stability of contaminants. These processes include sediment bioturbation and bioaccumulation. The former generally causes sediment oxidation, increasing degradation rate of organic contaminants. Bioaccumulation occurs in organisms when the absorption rate is greater than the metabolic removal, causing storage of the contaminant in the body of the organism. Biomagnification occurs in upper parts of the food chain when the contaminant is ingested by consumption.

14.6.2   GWF Comparison by Different Land Uses in a Hydrologic Basin

Diffuse sources of pollutants from runoff of soils with different use generally represent the largest pollutant source in hydrologic basins. Examples include the contributions of nitrogen and phosphorus from agricultural soils and gardens. Apart from the use of fertilizers, nutrients in food for humans and animals are imported to basins; pollutants generated outside basins can be deposited through atmospheric transport and deposition and it has been found that this mode may represent the biggest nitrogen source in hydrologic basins (Zhu et al. 2008).

Changes in land use affect not only polluting sources but also hydrology. In jungle basins, rainfall infiltrates through soil, water is lost by evapotranspiration or aquifers are recharged, which then feed rivers and streams. However, in agricultural catchments, surface runoff predominates. In urban basins, surface runoff is even more important due to the existence of impervious surfaces, reduced evapotranspiration and storm drainage systems, which carry the runoff water to rivers and streams. This situation also alters pollutant export due to the lack of biogeochemical attenuation and the related contaminant retention. The runoff path determines the contact time between pollutants, vegetation and soil, where biogeochemical processes minimize or remove them. When the rate is high, the contact time is generally low and contaminants are transported to rivers and streams with minimal removal.

14.6.3   GWF Application to Describe Advantages and Disadvantages of Transfers Among Hydrological Basins

Water deficit can be caused by natural events or human activities. Specific geographical or hydrological conditions can prevent water supply to meet demand. When water is overexploited to the point that it is not recharged in sufficient quantity or quality and when it is interrupted or has an inequitable distribution among users, it can cause social conflicts. Water deficit may result from overexploitation of the upper parts of hydrologic basins due to the construction of dams, water transfers to other basins and extraction for irrigation, thus altering downstream natural flow of rivers and water availability in quantity and quality (Pittock et al. 2009). Water quality degradation, expertise, and user conflicts can also cause water deficit situations (UN-Water 2007).

It is important to consider that water footprint of a city includes water supply infrastructure, drainage and wastewater treatment and transfer systems between hydrological basins (Engel et al. 2011).

14.6.4   GWF as Warning System of Pollution of Water Bodies

GWF values in a hydrologic basin indicate the violation degree of the criteria or limits of water quality. UNEP/CEOWM (2011) mention that indicators describing economic costs of gray water footprint have not yet been developed.

However, Hansen et al. (2007) developed a tool to facilitate the water value estimate in Lake Chapala under various scenarios, based on modules that describe water behavior in volume and quality, ecological values, and production.

It contains a hydrological component that allows to create scenarios of lake water extraction, reaching different equilibrium volumes (van Afferden/Hansen 2004). The environmental component considers environmental services as the lake capacity to assimilate pollutants, water pollution effects on its purification costs to supply the suburban area of Guadalajara, the loss of agricultural soil because of salinization when irrigating with lake water with high salt concentration, and degradation of forests and soils in the basin of the lake when the water level decreases.

14.7 Conclusion

GWF definition was revised and new applications of this tool to describe water pollution were proposed. It is proposed to include reactive transport processes and natural attenuation for better assessment of GWF of pollutant loads; it is suggested to use this concept in land use planning, which reduces GWF; its use is considered to describe advantages and disadvantages of transfers among hydrologic basins and, finally its use is proposed as a warning system of pollution of water bodies.

References

Chapman, Deborah. *Water Quality Assessments–A Guide to Use of Biota, Sediments and Water in Environmental Monitoring*. London: on Behalf of WHO by F & FN Spon, London: 1996.

CONAGUA. *Programa Nacional Hídrico 2007–2012*. México: SEMARNAT, 2008.

De Onis, Mercedes et al. "Estimates of global prevalence of childhood underweight in 1990 and 2015", *Journal of the American Medical Association* 291 (2004): 2600–2606.

Dourte, Dourte and Clyde Fraisse. *What is a water footprint?: An overview and applications in agriculture*, University of Florida IFAS Extension, 2012.

Engel, Katalina et al. *Big Cities. Big Water. Big Challenges. Water in an Urbanizing World*. Berlin: World Wilde Fund for Nature, 2011.

Hansen, Anne M. et al. *El Lago de Chapala y su entorno ecosocial. Desarrollo de una matriz de contabilidad social para el análisis de las políticas ambientales*, México: SEMARNAT-2002-CO1-0087, 2007.

Hansen, M. Anne and Carlos Corzo. "Evaluation of the Pollution of Hydrological River Basins", in *Water Resources in Mexico: Scarcity, Degradation, Stress, Conflicts, Management, and Policy*, Hexagon Series on Human and Environmental Security and Peace, edited by Ursula Oswald Spring, 201-2015. Berlin: Springer-Verlag, 2011.

Hoekstra, Arjen and Ashok Chapagain. *Globalization of Water: Sharing the Planet's Freshwater Resources*. Oxford: Blackwell Publishing, 2008.

Hoekstra, Arjen Ashok Chapagain and Mesfin Mekonnen. *The water foot-print assessment manual: setting the global standard*. London: Earthscan, 2011.

Liu, Cheng et al. "Past and future trends in gray water footprints of anthropogenic nitrogen and phosphorus inputs to major world rivers", *Ecological Indicators*, (2012): 42–49.

Mauch, Corine, Emmanuel Reynard, and Adhele Thorens. *Historical Profile of Water Regime in Switzerland (1870–2000)*, Working paper de l'IDHEAP Online-Ressource, III, (2000): 65 Accessed August 13, 2013.

Oswald-Spring, Úrsula. "Water Resources in Mexico: A Conceptual Introduction", in *Water Resources in Mexico: Scarcity, Degradation, Stress, Conflicts, Management, and Policy*, Hexagon series on Human and Environmental Security and Peace 7, edited by Ursula Oswald Spring, 5–17, Berlin: Springer-Verlag, 2011.

Pittock, Jamie et al. *Interbasin water transfers and water scarcity in a changing world—a solution or a pipedream?* Germany: WWF, 2009.

Prüss-Ustün, Annette et al. "Knowns and unknowns on burden of disease due to chemicals: a systematic review", *Environ Health* (2011): 10:9.

SEGOB. *Ley de Aguas Nacionales*, México: Diario Oficial de la Federación, SEGOB, 2004. Last amendment published on June 8 (2012): 1–106.

The United Nations Children's Fund/World Health Organization. *Diarrhea: Why Children are Still Dying and What Can be Done*. New York/Geneva, UNICEF and WHO, 2009.

The United Nations Children's Fund/World Health Organization. *Joint Monitoring Programme for Water Supply and Sanitation: Progress on Drinking Water and Sanitation*, New York/Geneva: UNICEF/WHO, 2010.

United Nations Environment Programme/The Chief Executive Office Water Mandate. *Water Footprint and Corporate Water Accounting for Resource Efficiency.* Oakland: United Nations Environment Programme/UNEP/CEOWM, 2011.

UN-Water. *Coping with water scarcity. Challenge of the twenty-first century.* UN-Water/ FAO. 2007.

United States Geological Survey (USGS) (2012), *Toxic Substances Hydrology Program, Natural Attenuation, Definitions,* http://toxics.usgs.gov/definitions/natural_attenuation.html, Accessed July 30, 2012.

Van Afferden, Manfred and Ann M. Hansen. "Forecast of lake volume and salt concentration in Lake Chapala, Mexico", *Aquatic Sciences* (2004): 257–265.

Zhu, Guibing et al. "Biological nitrogen removal from wastewater", *Reviews Environmental Contamination and Toxicology*, (2008): 159–195.

Chapter 15
Gray Footprint and Mining: Impact of Metal Extraction on Water

Germán Santacruz-de León and Francisco J. Peña-de Paz

Abstract Mining comprises the development phase, start-up and stripping activities for surface mining; it requires the construction of access roads, the work construction of water supply and electricity. In the first phase, the operation stage includes mineral extraction; the second phase involves processes of benefit and disposal of liquid and solid waste. The last stage involves the restoration and rehabilitation of the site. An underground mine design comprises three aspects: development, preparation, and exploitation; this type of mining allows to exploit seams that lie beneath the surface. It leads to lower noise emissions and dust is limited to the externally generated. In contrast, it requires greater technical complexities, it is more complicated, costly, and dangerous for the miners, so there is the tendency to abandon underground mining and to prefer open-pit mining (Buitelaar 2001). Currently, the largest mining activity takes place in the north-central region of Mexico; some estimates calculate that metal mining uses 53.5 million m^3 (Mm^3) of water, from surface or underground sources (López et al. 2001) and the volume of wastewater generated is estimated at 26.2 Mm^3, which is poured into water bodies or municipal drainage networks. Thus, mining affects quality and quantity of the liquid. Acid mine drainage is present in underground and open-pit mining and it is not only present in operating mines, but also after their closure. It is considered as the most serious and persistent mining environmental problem.

Keywords Gray water footprint · Mining · Semi-arid Mexico

15.1 Presentation

Because of the lack of reliable and detailed data on direct and indirect water consumption in the mining extraction processes and because of the diversity of extracted materials and of the processes used to do it, currently, it is not possible to establish the magnitude of the mining water footprint in terms of profit and volume consumed by the extraction process and—this is very important—the polluted effluent left by it and the material deposits that continue to affect aquifers and surface bodies, for many years even after mining companies have ceased operations.

© The Author(s) 2016
R.H. Pérez-Espejo et al. (eds.), *Water, Food and Welfare*,
SpringerBriefs in Environment, Security, Development and Peace 23,
DOI 10.1007/978-3-319-28824-6_15

As discussed above, this work is exploratory and methodological. We propose an approach to water footprint defined as the amount of water used directly and indirectly for a given type of mining, precious metals, and we suggest a methodology that not only considers water as an input in the production process to achieve metal separation, but also it incorporates medium- and long-term effects that this extractive activity has due to geomorphological changes it causes and that directly impact on the basin hydrological regime, both in terms of runoff types (direction, speed, permeability, etc.) and water quality, because wastes are added or removed during extraction and processing of metals. Both elements increase significantly the water volume used by metal mining.

After providing a framework for identifying the economic importance of this mining type, we will focus on an extractive mode, open-pit mining, because it is the dominant trend due to its profitability. Using the example of Minera San Xavier (MSX) in San Luis Potosi, we approach the impacts that an extraction mode has on water uses. After a schematic overview of the parts involved in mining extractive system, in the first part the existing statistics are analyzed in relation to mining in northern Mexico. In the second part, environmental and social impacts of mining are analyzed. Finally, it is made an inventory of the type of effects that mining has on water availability.

15.2 Metal Extraction Activities

Mining comprises the development phase, start-up, which covers the preparation of shafts and tunnels in underground mines, and stripping activities for surface mining: construction of access roads, work construction of water and electricity supply. In the first phase, the operation stage includes mineral extraction; the second phase involves processes of benefit and disposal of liquid and solid waste. The last stage comprises, according to mining manuals, the restoration and rehabilitation of the site (Anonymous, s.a.; Jiménez et al. 2006). As shown in Fig. 15.1, currently in Mexico there are several places where social mobilization has given attention to the type of mining exploitation being done or planned and its environmental effects, mainly those on water.

According to the shape and location of the ore body, mining methods can be divided into four basic types: (1) underground mines, using tunnels and galleries; (2) surface mines by opencast; (3) drill holes and; (4) undersea mining or dredging (UNEP 1994).

Underground mining selection is based on the deposit characteristics, such as size and dimensions, distribution and mineral mechanical characteristics, economic benefit criteria, etc. (UNEP 1994). An underground mine design comprises three aspects: (1) Development, involving work for deposit access, (2) Preparation, which consists of dividing the deposit into blocks, and (3) Exploitation, which are mineral extraction works (Jiménez/Molina 2006). This type of mining allows to exploit

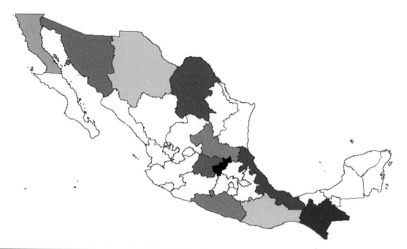

CONFLICT	STATE	PROJECT	COMMUNITY AFFECTED
Baja California says no to Paredones Amarillos of Vista Gold Corp.		Paredones Amarillos	Todos los Santos
Cananea, Group Mexico and Mining Union		Cananea	Inhabitants of Cananea
Communities in the municipality of Ocotlan claiming illegal mining concessions of Minera Natividad		Natividad Mine	Community of Calpulalpan de Mendez
Pasta de Conchos collapse		Pasta de Conchos Mine	
Ejidatarios rise against barite Chicomuselo Mine		Minera Caracol, Chicomuselo	Chicomuselo. Ejidatarios of Grecia, Monte Sinai, Nueva Morelia
The spill of Minera María		Minera María	
Great Panther contaminates dams		Cata Mine	Inhabitants of Cata
Oro Nacional Mine (Canadian) vs inhabitants of Mulatos, Sonora, Mexico		Oro Nacional Mine	Mulatos
Mineinders usurped lands from Ejidatarios of Huizopa		Minera Dolores	Ejidatarios of Huizopa
Minera contaminated water with arsenic in Cocula		Cocula	Indigenous people of Tlamacazapa
Minera San Javier operates outside the law		Cerro San Pedro	Ejidatarios of the Cerro San Pedro
Motozintla municipality, Chiapas, opposes Oro Mine in Ejido Carrizal		Motozintla	Ejidatarios of Grecia, Ejido Carrizal
Opposition to antimony plant in Queretaro		Flotation Plant	San Antonio de la Cal
La Luz project threatens to destroy the "Cradle of the Sun" for the Huicholes		La Luz	Real del Catorce
Veracruz opposes gold project within 3 kilometers of Nuclear Power Plant		Caballo Blanco	Veracruz

Fig. 15.1 Distribution of social conflicts by impacts of recent mining activity in Mexico. *Source* The Observatory of Mining Conflicts in Latin America, OCMAL, 2010

seams that lie beneath the surface; underground mining methods are generally classified into naturally supported cameras, artificially supported cameras and sinking (Gratzfeld 2004; Jiménez/Molina 2006; UNEP 1994).

Underground mining causes lower noise emissions and dust is limited to the externally generated. In contrast, it comprises advanced technology and skilled workers; it requires greater technical complexities and it involves high risks for workers. Thus, underground mining is more complicated, costly, and dangerous for miners (McMahon/Remy 2003). Because of this, it is preferred to use any of the superficial methods whenever it is possible; so there is a tendency to abandon underground mining and to prefer open-pit mining (Buitelaar 2001).

With technological advances, open-pit mining started to be employed more often. Surface mining is done by advancing horizontally on land cover, and it is called by different names depending on the type of extracted material: *open-pit mines* for metals; *open-cast mine* for coal or lignite; *quarries* for construction and industrial materials (sand, granite, slate, marble, gravel, clay, limestone, shale, quartz, talc, phosphate, salt, potassium, sulfur, etc.); and *pleasure mines* for heavy metals (gold and silver, platinum, iron, chromium, titanium, copper, tin, lead, zinc, etc.) and minerals (Gratzfeld 2004).

Open-pit mines consist of terraced pits, deep and wide, which regularly have a circular shape; extraction starts with the drilling and dynamiting of rock that, once classified, it is transported to the processing plants. In pleasure mines, low-compacted deposits of sand, gravel, silt, or clay are exploited, separating precious metals from them by sieves and laundries; they tend to be located in riverbeds or near them (Anonymous, s.a.; Matamoros and Vargas 2000).

15.3 Gold Mining Contribution to National Economy

In the first decade of the twenty-first century, the high mineral prices remained, stimulating an investment increase in exploration. Globally, 10,500 million dollars were invested in this task, exceeding by 40 % the expended in 2006. In Latin America, the main recipients of that investment were Mexico, Peru and Chile, which together received 24 % of it. In the same year, 4,410 million dollars were allocated for gold exploration (Anonymous 2008).

Currently, the largest mining activity takes place in the north-central region of Mexico (Fig. 15.2). The main mining centers are located in the states of Sonora, which is the largest producer of gold and copper; Coahuila, main producer of antimony and bismuth; Zacatecas, first in production of silver; Chihuahua, which is a leading producer of cadmium, zinc, and the only one with tungsten deposits. Baja California Sur, San Luis Potosi, and Durango are also noteworthy; they are states where significant metal deposits have been located (Coastal 2008).

The figures are revealing. In 2000, the mining sector contributed between 1.17 and 1.5 % to the gross domestic product (GDP) and it participated with 1.5 % of national employment. On the other hand, the large-scale mining generated 84.1 % of the total value of domestic mining–metallurgical production; medium and small mining contributed 13 and 2.9 %, respectively (Center for Competitiveness Studies 2004).

Fig. 15.2 National position of producing states of conceivable minerals* in northern Mexico, 2007. *Source* based on data from Anonymous 2007. *Minerals with this name, according to the Mining Law of Mexico, are those that can be exploited only with permission or concession granted by the Secretariat of Economy

In 2005, Mexico ranked first in silver and celestite production; it was ranked among the top five producers of cadmium, arsenic, and bismuth. It is also among the ten largest producers of gold, manganese, and antimony. In the same year, mining–metallurgical production amounted to 53,954 million pesos; the states of Sonora, Zacatecas, Coahuila, Durango, San Luis Potosi, and Chihuahua stood out for their production value (Mining Chamber of Mexico 2006); however, other figures show that the value of mineral production in 2005 totaled 71,800 million pesos (Anonymous 2007). This contributed 1.6 % to GDP. 912 million dollars were invested and 120 million dollars were spent for exploration of new deposits (Mining Chamber of Mexico 2006; Jiménez et al. 2006).

In 2007, gold production amounted to 72,600 kg (Fig. 15.3), which is the most dynamic production due to investment flows for opening new mines and to high international prices.

The investment increment or decrement is reflected in the number of jobs created. In 2007, people employed in mining increased to 292,993; however, in 2009 and 2010, the mining sector employed 269,501 and 283,800 people, respectively (Fig. 15.4). Mining of metallic minerals, which corresponds to branch 13, generated 6,543 of those jobs (Anonymous 2010).

The above comparison (Fig. 15.4) is done with the premise that the mineral mining corresponding to branch 11 is more environmentally friendly, and it generates more jobs, compared with the negative environmental impact resulting from metallic mineral mining.

Mining sector contributes 1.5–2.5 % to GDP; however, no official statistic indicates the environmental costs of that contribution. In recent years, mining has contributed 1.52–1.94 % of total national employment.

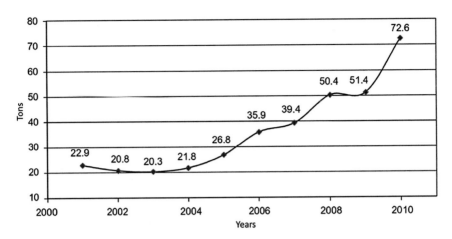

Fig. 15.3 Gold production (ton) in Mexico, 2001–2010. *Source* Anonymous, 2010

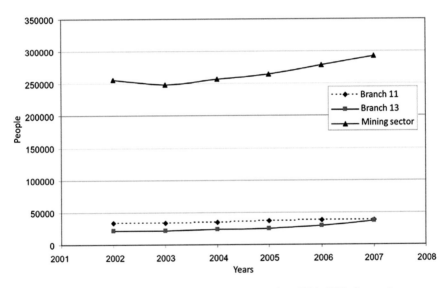

Fig. 15.4 Number of jobs created by mining sector in Mexico, 2001–2007. *Source* Anonymous 2007

For comparison, it can be seen that official statistics show that sand and gravel extraction—which is within the mineral extraction industry—generated at least from 2003 to 2005, similar economic wealth to the one produced by gold and silver extraction. In the period 2003–2010, the states of Sonora, Chihuahua, and Zacatecas account for most of the mining production and, thus, with the largest generation of economic resources of the branch. These three states, in conjunction with San Luis Potosi, Sinaloa, are the major gold producers in northern Mexico.

Table 15.1 Major operating mines of metallic minerals in San Luis Potosi, 2007

Name	Company	Municipality	Mineral
Cerro de San Pedro	Minera San Xavier	Cerro de San Pedro	Au, Ag
El Rey-Reyna	Industrial Minera Mexico	Charcas	Au, Ag, Zn, Pb
San Acasio y Pilar	Minera Santa María de la Paz	La paz	Au, Ag, Cu

Source Based on data from SMG (2008d)

Next, the main characteristics of mining production in some of these states are indicated, with emphasis on the type of mining used.

In 2007, the state of San Luis Potosi occupied nationally the fifth place in the production of metallic minerals, and fourth place as gold producer. This state has great potential of metallic minerals such as gold, silver, copper, lead, zinc, manganese, tin, iron, mercury, and antimony (SGM 2008c). It has three active mines, in each of them gold and other metals such as silver, lead, zinc, and copper are extracted (Table 15.1 and Annex).

15.4   Mining Environmental Impact: Effects on Water Availability

In mining, the most important stage is the metallic mineral extraction, which in the mining slang it is known as profit. The mined rock contains valuable components of economic interest and sterile components—known as bargains—that regularly have no economic value (Anonymous, s.a.). Mineral processing can be simple or may involve complex processes; this activity can be done on site or can be carried out elsewhere. Whatever the condition, it still implies significant amounts of water (Fig. 15.5), which can be difficult to access in arid and semi-arid areas (Gratzfeld 2004).

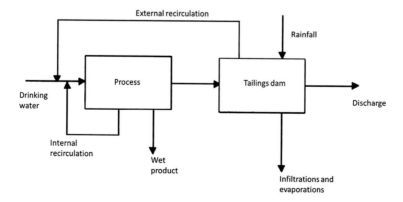

Fig. 15.5 General flow diagram of water use in the mining industry. *Source* Taken from Rao and Finch (1988), quoted in Pacheco and Duran (2006)

Table 15.2 Water use in metal mining in Mexico

Mining branch	Extraction (m³/year)	Recirculation (m³/year)	Demand (m³/year)	Consumption (m³/year)	Descharge (m³/year)
Precious metals	25,632,534	20,511,451	46,144,015	11,676,303	13,956,232
Nonferrous metals	6,810,026	16,041,354	22,851,350	2,685,910	4,124,115
Steel minerals	21,149,833	37,107,785	58,257,618	12,991,538	8,168,298
Total (m³/year)	53,592,393	73,660,590	127,252,983	27,353,751	26,248,645
Equivalent population	734,142	1,009,049	1,743,192	374,709	359,570

Source Modified from López et al. (2001)

Some estimates calculate that Mexican metal mining uses 53.5 million m³ (Mm³) of water, from surface or underground sources (López et al. 2001). This volume would be sufficient to provide 200 l of water per day for one year to a population of 734,000 inhabitants. The volume of wastewater generated is estimated at 26.2 Mm³ (Table 15.2), which is poured into water bodies or municipal drainage networks.

In the various activities involved in mining of metallic materials—extraction and processing—there are adverse impacts on water resources where the mine is located, but usually occur differentially (Calva 1994). Thus, mining affects quality and quantity of the liquid. Benefit plants and tailings dams, where the mining waste is dumped, are a source of environmental pollution; in these plants, metals of interest are separated from the rocks, so the amount of wastes generated and the degree of contamination depend on the mineralogical composition of the mine and the benefit technique (Jiménez et al. 2006). In accordance with EIM, in the Cerro de San Pedro mine project, located in San Luis Potosi, one million cubic meters of water per year are required (Santacruz 2008), following the procedure of equivalent population, this volume would provide 200 l per capita per day to a population of 13,697 inhabitants. MSX originally proposed the use of treated wastewater; however, it has applied for the authorization to acquire water rights for agricultural use to the National Water Commission (Conagua in Spanish). Thus, one million cubic meters will be extracted from the aquifer from which the water is obtained to meet the liquid needs of the city of San Luis Potosi. Most of the water will be used in leaching pads and dust suppression systems; MSX states that there may be an extraordinary water consumption, which would increase to 1.3 Mm³.

To meet the water needs of the city of San Luis Potosi, the intermunicipal operating organism has allocated 85 Mm³/year (Peña 2006). There is a recharge of 78.1 Mm³/year, although there are controversies about which part of the system aquifer receives this recharge. Currently, it is said that this volume feeds the shallow aquifer; in general, 149.34 Mm³/year is allocated, implying a deficit of 71.4 Mm³/year (Conagua 2002). During the time period covered by the MSX project, the extracted volume for mining purposes will be added to the latter. Since 1961, the low water availability in San Luis Potosi caused the prohibition of the aquifer for any purpose other than domestic use; groundwater extraction is considered, by MSX in its *Environmental Impact Manifestation*, as significant adverse.

Water can be a mining input; but in many cases, it can be seen as a problem by mining companies.

Underground mining causes less visible effects, but no less harmful to the environment than open-pit mining. It causes aquifer abatement because of the continuous water pumping from inside the mine (Coll-Hurtado et al. 2002); in the mining slang, wasted water is known as acid mine drainage (AMD), which, if not adequately treated, will contaminate the soil or water bodies where it is poured; this reduces its quality for human consumption or agricultural use.

Acid mine drainage is present in underground and open-pit mining. It is caused by the oxidation of minerals containing sulfide, producing sulfuric acid (H_2SO_4) with pH between 1.5 and 7; thus, this substance can dilute easily metals such as iron, cadmium, copper, aluminum, and lead (Fernández 2008; Gratzfeld 2004; Jiménez et al. 2006). AMD can be defined as the inorganic chemical pollution of water, resulting from the oxidation of minerals containing sulfide (UNEP 1994).

AMD is not only present in operating mines, but also after their closure. It is considered the most serious and persistent mining environmental problem; although it occurs in most mining exploitations, it is magnified in areas where rainfall is considerable (Anonymous 2002).

The discharge of acid drainage into water bodies causes their pollution and thus, the incorporation of metals in the food chain; it can also pollute the aquifers, causing the water contained in them to be inadequate for human consumption. Contamination incidents are serious when acid drainage, stored in underground abandoned mines, contaminates aquifers that are the source of water for domestic consumption (Anonymous 2002; Tovar, s.a.).

Regarding the AMD, the Minera San Xavier case can be mentioned, which states that in the operation stage, in the realization of the pit, and according to information from more than 200 exploration boreholes, the water table of the region will not be not intersected until the maximum depth of the planned pit (Santacruz 2008). However, in the *Environmental Impact Manifestation*, it indicates that: "As the pit development proceeds and particularly towards the final stages of the same, it is expected that some outcrops of potentially acid generating rock will be exposed to oxidation with the consequent possibility that it will help in the generation of acidic pH solutions and it can contain metals in solution and dissolved and suspended solids" (Behre Dolbear 1997:328).

In addition, Carrillo (2005), quoting Alloway (1995), indicates that the elements associated with gold extraction are silver, tellurium, arsenic, antimony, mercury, and selenium; and in the case of silver, elements such as copper, antimony, lead, tellurium, and zinc are associated. In that sense, the EIM of MSX mentions that of the 117 million tons of dump, about 600,000 tons of material known as intrusive porphyry with sulfide will go to the dumps, "with the risk of generating, in the long-term and during the period of total sulfide oxidation, solutions with an acid pH that may contain metals in solution and affect aquifers [sic] of the region" (Behre Dolbear 199:328).

15.5   Conclusions

During the last decade, concessions and projects to open new mining extraction of metals, particularly gold and silver, have increased in the central and southern Mexico. These projects have brought the mobilization of multiple communities whose environmental security and, in particular, that concerning the availability of good quality water is threatened.

The North is where the principal mines are located, especially those intended for the metallic mineral extraction. Currently, to increase the profitability of several of the sites historically engaged in the extraction, the open-pit mining method is chosen, being the method that generates greater negative environmental impact. Technological advances—the use of sophisticated equipment for exploration and the use of metallurgical processes such as cyanidation that allow the extraction porphyry metal deposits—have allowed open-pit mining to be the most used by mining companies.

Although these methods are presented as methods that consume less water due to the reuse of cyanidation systems for metal separation, open-pit mining has a greater geomorphological impact because it modifies large surfaces to extract the mineral. Changes of this magnitude also affect runoff type, speed, recharge capacity, direction of surface currents and, especially, it threatens water quality because large amounts of waste are kept without proper management.

On the one hand, mining water footprint calculation requires public and accurate record of the volumes used directly and indirectly in the processes of obtaining the mineral and its benefit. But above all, it requires special attention to changes affecting availability of good quality water and the disturbance of basins and subbasins where such types of companies are established. Mining is a typical case of expanded use of water concessions received. They are extended in space and time, because their effects on watercourses and, thus, on the water they use, remain beyond the extraction period (via pollution, for example) and amplify their influence in space due to the intervention on the geomorphological basin configuration.

Annex: Extraction Yields of Various Mining Products by Federal State

State	2007	2008	2009	2010	2011
Baja California					
Gold (kg)	–	–	–	358	645.8
Silver (kg)	–	–	–	14284.00	10920.00
Aggregates	4476043.00	634737.00	16421163.50	17642572.35	12840479.10
Clays	66456.00	66456.00	46800.00	45000.00	46000.00
Sand	1836109.90	1329054.06	21299489.84	21285569.84	18857962.50

(continued)

(continued)

State	2007	2008	2009	2010	2011
Limestone	354432.00	1025992.80	249600.00	240000.00	240000.00
Gravel	1358726.10	868862.80	481728.00	463200.00	463200.00
Gypsum	22152.00	43336.00	15600.00	15000.00	15000.00
Chihuahua					
Gold	12891.30	13140.60	15221.80	18256.60	15262.30
Silver	451292.00	466242.00	580271.00	783081.00	794238.00
Cadmium	362.45	350.03	341	–	–
Copper	13633.00	13914.00	13433.00	13132.00	12468.00
Iron	378213.00	381661.00	106807.00	438421.00	212399.00
Lead	58657.00	56235.00	53169.00	46308.00	47053.00
Zinc	136437.00	142035.00	150211.00	133734.00	122254.00
Aggregates	5300.00	27020.00	48750.00	41285.00	640529.00
Clays	30878358.00	31115142.75	1036691.50	945019.00	640529.00
Sand	3686715.00	2366234.00	3403784.00	2849376.00	3009015.00
Barite	–	–	–	850	600
Limestone	3623393.00	5208263.00	3048784.00	4142418.00	1582704.00
Kaolin	72000.00	107005.00	61500.00	106000.00	106500.00
Dolomite	–	–	–	6001.00	4771.00
Gravel	4020182.40	3021217.00	4430550.00	4767639.00	3885050.00
Perlite	365	180	31	29	–
Slate	518035.00	437581.00	400000.00	447593.00	388222.00
Dimensionable rocks	11140.00	55700.00	8570.00	9678.00	8450.00
Salt	3000.00	7500.00	7930.81	5450.00	4320.00
Gypsum	156000.00	157304.25	120800.00	168000.00	138050.00
Zeolite	200	–	–	–	–
Coahuila					
Gold (kg)	1.1	1.2	0.2	–	0.1
Plata (kg)	35134.00	41988.00	38860.00	122602.00	134452.00
Antimony	414	380	74	71	5
Bismuth	1170.00	1132.00	854	863	875
Cadmium	–	–	–	863	875.64
Copper	9	9	2	–	–
Tin	19	15	–	–	–
Iron	3233568.00	3838719.00	5179379.00	4595325.00	3601546.00
Lead	568	1340.00	1154.00	964	30
Zinc	–	4	–	–	–
Aggregates	–	1233966.86	618696.00	629927.00	851631.00
Sand	3907928.00	3204500.00	2798574.00	3082474.00	2436000.00
Barite	29977.00	26265.00	30675.00	22161.00	28023.00

(continued)

(continued)

State	2007	2008	2009	2010	2011
Bentonite	47000.00	23500.00	40000.00	15000.00	–
Calcite	16000.00	8000.00	96000.00	84000.00	72000.00
Limestone	3821178.00	5654069.00	6816231.00	5505391.00	2778084.00
Coal	17299221.00	10402658.00	9496189.00	11246639.00	13718159.00
Celestite	96902.00	29621.00	36127.00	31429.00	40699.00
Dolomite	760079.00	813812.00	781398.00	1161069.00	2462119.00
Fluorite	133578.00	139429.00	108930.00	121833.00	119516.00
Gravel	5201658.80	4265300.00	3724296.00	3727381.00	3242400.00
Dimensionable rocks	194735.00	424143.70	506328.70	726328.70	1200.00
Salt	620000.00	18261.00	19309.93	31761.00	–
Silica	736100.00	738467.00	777863.00	814591.00	760940.00
Magnesium sulfate	33900.00	43053.00	34700.00	39400.00	45598.00
Sodium Sulfate	605000.00	618000.00	606000.00	620000.00	630500.00
Gypsum	348447.50	400653.00	270031.00	299113.00	276216.00
San Luis Potosi					
Gold (kg)	1689.00	3588.60	4346.90	4794.50	5619.00
Silver (kg)	109068.00	135123.00	152441.00	179895.00	162084.00
Arsenic	513	–	–	–	–
Cadmium	–	–	–	600.57	609.37
Copper	20198.00	19742.00	19907.00	21632.00	21128.00
Iron	–	–	–	–	693
Lead	3534.00	5608.00	5210.00	4189.00	3736.00
Zinc	65610.00	63463.00	62463.00	58040.00	53489.00
Aggregates	120000.00	2600000.00	1350000.00	462000.00	910000.00
Clays	780090.00	850000.00	950000.00	923000.01	1115000.00
Sand	6628669.00	7492040.00	8398200.00	7777020.00	5405100.00
Bentonite	4800.00	5100.00	6000.00	5800.00	6800.00
Calcite	326016.00	193950.00	197600.00	178600.00	188100.00
Limestone	4160480.00	4462310.00	6375200.00	4802800.00	4675200.00
Quarry	2	21728.00	24600.00	15400.00	17635.00
Kaolin	1760.00	3200.00	3600.00	6300.00	7300.00
Fluorite	799783.00	918220.00	937010.00	945553.00	1087391.00
Phosphorite	6 000.00	–	–	–	–
Gravel	10067726.40	13024000.00	14182020.00	13094500.00	9251300.00
Dimensionable rocks	110560.00	51500.00	48900.00	53000.00	53000.00
Salt	100000.00	8000.00	8459.53	8000.00	8000.00
Silica	31189.00	33657.00	32253.00	34727.00	41682.00
Tepetate	1200000.00	600000.00	3500.00	3500.00	3500.00

(continued)

(continued)

State	2007	2008	2009	2010	2011
Tezontle	1200000.00	600000.00	600000.00	10000.00	2500.00
Gypsum	260030.00	427000.00	1362213.00	461200.00	287756.00

References

Asociación de Ecologistas Costarricence-Amigos de la Tierra, *Minería de cielo abierto y sus impactos ambientales*, prepared for the Frente Nacional de Oposición a la minería de Oro a Cielo Abierto, Asociación de Ecologistas Costarricence-Amigos de la Tierra, Costa Rica, 2001.

Anonymous. *Breaking New Ground: Mining, Minerals and Sustainable Development, MMSD Project Report*. London: International Institute for Environmental and Development-World Business Council for Sustainable Development, 2002.

Anonymous. "Capítulo 1. Resumen de indicadores básicos de la minería." In *Anuario Estadístico de la Minería Mexicana 2013*, 10–40. México: Servicio Geológico Mexicano-Secretaría de Economía, 2007.

Anonymous. *Manual de minería*. Lima, Perú: Mining studies of Peru, s.a.

Behre, Dolbear. *Manifestación de impacto ambiental,* Requested by Minera San Xavier. Jalisco, Mexico: Behre Dolbear of Mexico, 1997.

Buitelaar M., Rudolf. A*glomeraciones mineras y desarrollo local en América Latina*. Bogotá: Centro Internacional de Investigaciones para el desarrollo, CEPAL/Alfaomega, 2011.

Cámara Minera de Mexico. *La industria minera en México*. México: Cámara Minera de México, 2006.

Calva, Héctor. *La minería mexicana en el siglo XX*. México: Fondo de Cultura Económica, 1994.

Carrillo, Rogelio. "Breve descripción de la minería en México". In *El sistema Planta-Microorganismo-Suelo en áreas contaminadas con residuos de minas* edited by Ma. del Carmen Gonzalez, Jesus Pérez and Rogelio Carrillo. 137–153. México: Colegio de Posgraduados, 2005.

Carrillo, Joel and Antonio Cardona. "Entorno hidrogeológico de San Luis Potosí". In *Agua subterránea,* Price, M., 274–320. México: Limusa-Noriega, 2003.

Centro de Estudios de Competitividad. *El sector minero en México: diagnóstico, prospectiva y estrategia"*. México: Instituto Tecnológico Autónomo de México, 2004.

Coll-Hurtado Atlántida, María Teresa Sánchez-Salazar and Josefina Morales. *La minería en México*. Col. Temas Selectos de Geografía de México (I.5.2). México: Instituto de Geografía, UNAM, 2002.

Costero, Cecilia. "Minera San Xavier, San Luis Potosí. Un estudio desde un punto de vista internacional". In *Internacionalización económica, historia y conflicto ambiental en la minería. El caso de Minera San Xavier*, coordinated by Cecilia Costero, 59–103, San Luis Potosí: El Colegio de San Luis Potosí, 2008.

Estrada, Adriana. *Impactos de la inversión minera canadiense en México: una primera aproximación*, México: Fundar-Centro de Análisis e Investigación, 2001.

Fernández, Juan Carlos. "Una aproximación al conocimiento del impacto ambiental de la minería en la Faja Pirítica Ibérica", *Revista de la Sociedad Española de Mineralogía (Spanish Mineralogical Society Bulletin)* 10 (2008): 24–28.

González, Silvia and Marta Sahores. *Impacto Ambiental debido al uso del Cianuro en la Minería a Cielo Abierto*, School of Natural Sciences, National University of the Patagonia San Juan Bosco, Argentina.

González, Jenaro. *Minería y riqueza minera de México*, Monografías Industriales del Banco de México, México, 1970.

Gratzfeld, Joachim. *Extractive industries in arid and semi-arid zones: environmental planning and management*, International Union for the Conservation of Nature and Natural Resources, Gland, Switzerland and Cambridge, United Kingdom, 2004.

Herrera-Canales, Inés. *La minería mexicana de la Colonia al siglo XX*, México: Consejo Nacional de Ciencia y Tecnología, 1998.

Jiménez, Carolina, Pilar Huante and Emmanuel Rincón. *Restauración de minas superficiales en México*, México: Secretaría de Medio Ambiente y Recursos Naturales, 2006.

Jiménez, Indhira and Jorge Molina. "Propuesta de medición de la productividad en minería de oro vetiforme y reconocimiento de estándares productivos sostenibles", *Boletín de Ciencias de la Tierra* 19, Medellin, Colombia. (2006): 73–86.

McMahon, Gary and Remy, Félix. *Grandes minas y la comunidad*. Bogotá: International Development Research Centre, World Bank, Alfaomega, 2003.

Santacruz de León, Germán. "La minería de oro como problema ambiental: el caso de Minera San Xavier". In *Internacionalización económica, historia y conflicto ambiental en la minería. El caso de Minera San Xavier*, coordinated by Maria Cecilia Costero Garbarino, 103–122. México: El Colegio de San Luis, 2008.

Servicio Geologico Mexicano. *Resumen de indicadores básicos de la minería*. México: Secretaría de Economía, 2007.

UNEP. *Human Development Report*. Oxford: UNDP, Oxford University Press, 1994.

Chapter 16
Considerations on Virtual Water and Agri-food Trade

Thalia Hernández-Amezcua and Andrea Santos-Baca

Abstract Virtual water (VW) content refers to the volume of water used to produce a good or service that is susceptible to be traded internationally. A significant feature that characterizes external trade of the country is the fact that virtual water imports exceed exports (Arreguín 2007), which has resulted in a purely accidental circumstance, but whose discovery allows to consider that it is possible to formulate public policy incentives to guide the best use of water resources through better allocation of water and more accurate selection of productive practices. With the inclusion of agriculture in NAFTA, food production and consumption in Mexico were closely linked to the largest food producer worldwide; on the one hand, imports of cereals and oilseeds were favored, but on the other hand, the production and export of vegetables and some fruits changed water use and stress. VW volume imports increased by 78 %.

Keywords Virtual water · Agriculture · Food trade

16.1 Virtual Water and Free Trade

The concept of virtual water is one of the analytical and methodological developments that puts into perspective the importance of water in the process of promoting well-being. It refers to the water content required to produce a good or service susceptible to be traded internationally. The intuitive approach to this idea is that economic exchange takes place on the basis of specialization processes arising from the presence of Ricardian comparative advantages. Economies with a relative abundance of water supplies would tend to specialize in the production of goods and services having cost advantages derived from the proportional abundance of some factors. Recently, virtual water flows in foreign trade have begun to be documented for the Mexican economy.

A significant feature that characterizes external trade of the country is the fact that virtual water imports exceed exports (Arreguín 2007). In terms of its implications for economic policy and growth promotion, this has resulted in a purely accidental circumstance, but whose discovery allows to consider that it is possible

© The Author(s) 2016
R.H. Pérez-Espejo et al. (eds.), *Water, Food and Welfare*,
SpringerBriefs in Environment, Security, Development and Peace 23,
DOI 10.1007/978-3-319-28824-6_16

177

to formulate public policy incentives to guide the best use of water resources through better allocation of water and more accurate selection of productive practices.

In a general sense, the prospect of virtual water (VW) allows to identify optimal production sites, in terms of the theory of comparative advantages, to meet the needs with the least possible pressure on ecosystems. The concept, as such, combines three dimensions: food security, free trade, and ecological vision.

Food dimension consists of knowing water efficiency in agro-alimentary of agri-food use taking into account availability and physical characteristics of the resource and territory. The proposed water footprint (WF) concept is to estimate water requirement for food production and exploit comparative advantages in order to reduce pressure on water resources.

From this idea, it follows that water flow in international food trade is the means to open domestic water markets and to ensure that this resource is used more efficiently like other productive resources. Thus, in theory, knowing the virtual water contained in food allows better management of water resources and the reorientation of production to areas where resource availability allows efficient and sustainable production (Renault 2002:3).

The pro-free market ideology obtained a new justification source and a renewed strength of implementation with the VW prospect, by adding to the theory of comparative advantage the thesis that the market promotes an interdependent system that tends to balance water resource disparities and to ensure its most efficient use. Thus, in terms of water, foods that flow from the most productive countries toward those less productive, generates two types of benefits: savings in water resources for the food importer country and savings in water resources globally (Renault 2002:15).

Undoubtedly, the introduction of the VW perspective was a breakthrough in the intention of achieving integrated water resources management by adding a physical dimension in terms of water volume to the analysis of monetary productivity. However, its close relation with free trade can lead to a partial view of how trade effectively organizes itself worldwide and therefore, the economic or water benefits that may arise from it.

In this chapter, a number of aspects to be considered are presented to critically qualify the approaches of the VW advantages, trade liberalization, and efficiency of the *invisible hand*.

- Potential savings of water incorporated into food imports only materializes if the decrease in food production releases water resources that may be available for other uses (Renault 2002:10);
- VW trade advantages should be considered in light of possible adverse and conflicting consequences for different actors in local economies. This leads to two questions: (1) who gets the political and security benefits of the VW flow? and (2) what consequences does this flux have in socioeconomic relationships within and between states (Warner 2003:125);

- 'Virtual' nature of water and free market make this resource management a non-political process. The nonpolitical character comprises an automatic-commercial water management and, therefore, non communitarian-participative (Warner 2003:130);
- The opening of national water markets through the integration to the global food market implies greater exposure and sensitivity of consumers and producers to the changes of international prices (Warner 2003:131). This has been demonstrated in the food prices increment in 2008 caused, mainly, by reasons external to the agri-food system, such as financial speculation and fuel market;
- It is important to recognize that the comparative advantages recognized in some countries are not the result of the "free play of market forces;" a known example is the agriculture of the United States and Europe, which function with high subsidies. Therefore, in many cases, the VW flow contained in foods of low price is based on trade distortions, subsidies, *dumping*, and other disloyal practices (Warner 2003:131). What appears to be water savings is only a mechanism of foreign market appropriation at the expense of local producers;
- For peripheral countries, world trade means participation in a system dominated by powerful transnational corporations. So that the shortage of food and water can be induced by large corporations, which can appropriate water resources and control large food inventories (Warner 2003:131).

Food trade based on the VW approach or comparative advantages displaces domestic producers by the reallocation of productive resources toward more efficient economic or hydrological sectors, generating harmful social effects such as South–North or field–city migration, poverty, and dependence on international food markets dominated by transnational corporations (Warner 2003:132).

In order to address water savings obtained through trade, the benefits of free trade and market are critically analyzed as automatic and efficient mechanisms of distribution and use of resources and social wealth.

16.2 Contemporary Agri-Food Regime in Mexico

According to Harriet Friedmann (1993:16), neoliberal food regime is characterized by the fact that it forces people and ecosystems to adjust to the markets. It arises with the implementation of world market as an organizing principle of food production and consumption and with the market opening to the largest agri-food companies in developed countries.

The current regime is based on a new international division of agricultural labor characterized by the specialized production of nontraditional units, called as "high value" in dependent countries, and mass production of cereals for export in developed countries, mainly the United States. For peripheral countries, this has meant increasing dependence on basic food imports. This "new agriculture" of the Southern countries involves the development of agri-food complexes, such as fruits and vegetables by contract, and inputs production (antibiotics, foods, genetics) in livestock (McMichael 1992:350). During 1970–1996, member countries of the Organisation for Economic Co-operation and Development (OECD) accounted for 76 % of cereal exports and peripheral countries for 60 % of imports (McMichael 2009:150).

This reconfiguration of the food market relationships was performed using two mechanisms of trade liberalization: debt regime and multi or bilateral trade agreements (Araghi 2009).

The case of Mexico is significant; by the end of 1982, the external debt increases, agricultural policy is restricted and the economy, as a whole, is closely linked to the cycles of macroeconomic stabilization and adjustment policies, as well as to external trade objectives. The national food programme 1982–1988 replaced the objective of "food self-sufficiency" by "food security," and thereafter, the policy objective was to obtain cheap food through imports (Appendini 2001:102–105).

The decisive step toward trade liberalization is taken with the North American Free Trade Agreement (NAFTA). While in the General Agreement on Tariffs and Trade (GATT) predominated the reluctance to include the agricultural sector in multilateral negotiations and the Uruguay Round remained stalled, the Mexican government decided to include it in this trilateral treaty. With NAFTA, a transformation of the productive structure of the sector is driven, modifying employment and land use, and generating a reallocation of these resources of activities unable to compete with imports from the United States and Canada, to export sectors (Puyana/Romero 2008:35). One argument was that the United States, being the main exporter in the world of agricultural products, could become the best and permanent supplier of food in Mexico.

With the inclusion of agriculture in NAFTA, the production and consumption of food in Mexico were closely linked to the world's largest food producer; on the one hand, imports of cereals and oilseeds were favored, and on the other, the production and export of vegetables and some fruits, which changed the use and water stress, as shown below.

16.3 Virtual Water in Agri-Food Trade Between Mexico and the United States

Food trade between Mexico and the United States grew significantly after the signing of NAFTA. Table 16.1 presents the total agricultural trade balance of Mexico and with NAFTA countries. From the total Mexican agricultural exports, 80 % goes to the United States and Canada. In relation to total imports, Mexico shows a more diversified pattern, as only 51 % comes from countries that are part of NAFTA. However, the ones of the agricultural sector are significantly concentrated in the NAFTA region, concentration that grew by more than seven percentage points with the entry into force of the treaty.

Table 16.1 Total and sectoral trade in Mexico, previous year to entry into force of NAFTA in 2010 (millions of dollars)

Trade		Exports				Imports			
		1994	Part. (%)	2010	Part. (%)	1994	Part. (%)	2010	Part. (%)
Total	Total	58,652	100	338,451	100	66,808	100	325,005	100
	NAFTA	48,858	83	283,991	84	47,676	71	170,672	53
Oil and gas	Total	7,229	100	41,086	100	1,570	100	24,058	100
	NAFTA	4,833	67	35,368	86	1,265	81	17,114	71
Nonfood industry	Total	47,798	100	279,901	100	59,325	100	280,015	100
	NAFTA	40,780	85	234,471	84	41,963	71	136,399	49
Agri-food	Total	3,625	100	17,464	100	5,913	100	20,932	100
	NAFTA	3,245	90	14,152	81	4,448	75	17,159	82

Source Secretariat of Economy with data from Bank of Mexico

To observe the behavior of the trade balance between both countries in a more disaggregated level, a classification for the main groups of exported (grains and legumes, fruits and vegetables, sugar, coffee, and processed products) and imported (staple food, fodder, meat, milk, oil, sugar, and others) products was made. Similarly, the analysis was divided into two study periods; the first of 1986–1993, which represents the period before NAFTA and the second period of 1994–2010, showing changes that occurred after it.

Exports from Mexico to the United States of selected food groups grew to 76 %, at an average annual rate of 9 % since the treaty; grains and legumes increased from 632 tons per year, in average, in the first period, to 6630 tons in the second, at an annual rate of 15 %; they were followed by processed products (11 %), fruits and vegetables (10 %), sugar (8 %), and coffee (3 %) (Fig. 16.1).

Imports increased 84 %, at an annual rate of 10 % from 1994 to 2010. Meat products, staple food, and feed stand out because they grew at an annual rate above average and represent the largest volume of imports from the United States. Sugar is also noteworthy, whose import was reduced at an annual rate of 1 % (Fig. 16.2).

To observe the implications of this change in the pattern of agricultural trade in Mexico in terms of VW, VW content is incorporated into each product group. Figure 16.3 show the VW trade between Mexico and the United States.

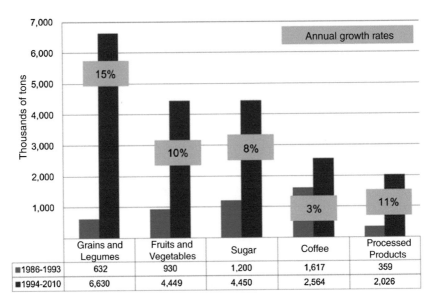

	Grains and Legumes	Fruits and Vegetables	Sugar	Coffee	Processed Products
■1986-1993	632	930	1,200	1,617	359
■1994-2010	6,630	4,449	4,450	2,564	2,026

Fig. 16.1 Mexico–United States exports by groups of agricultural products. Total exports (thousands of tons) by period. *Source* based on data from FAO (2012) and Hoekstra (2010)

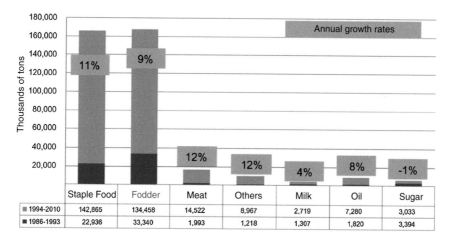

Fig. 16.2 Imports into Mexico from the United States by agricultural and livestock products. Total imports (thousands of tons) by period. *Source* based on data from FAO (2012) and Hoekstra (2010)

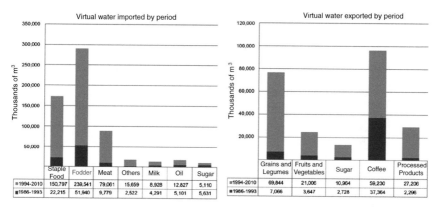

Fig. 16.3 Virtual water imported and exported by period. *Source* based on data from FAO (2012) and Hoekstra (2010)

As shown in Fig. 16.4, Mexico has always been an importer of VW in agri-food products. However, it is clear that after the forced specialization by NAFTA, Mexico became a country more dependent on American agri-food products. VW imports grew to 80 % compared with the previous study period, while volume imports increased to 78 %.

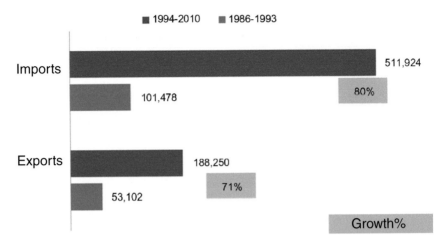

Fig. 16.4 Water balance Mexico–United States (m^3). *Source* The authors' elaboration with data from the Secretariat of Economy

References

* indicates internet link (URL) has not been working any longer on 8 February 2016.

AgroDer. *Huella hídrica en México en el contexto de Norteamérica.* México: WWF, AgroDer, 2012.

Appendini Kirsten. *De la milpa a los tortibonos, la restructuración de la política alimentaria en México.* México: El Colegio de México/United Nations Research Institute for Social Development, 2001.

Araghi, Farshad. "Peasants, Globalization, and Dispossession: A World Historical Perspective." Paper Presented at the Annual meeting of the American Sociological Association, San Francisco, California, August 7-11, 2009.

Arreguín, Felipe et al. "Agua virtual en México" *Ingeniería Hidráulica en México.* vol. XXII, no. 4 (2007): 121-132.

*Bilaterals, BIOTHAI, GRAIN. *Combatiendo los TLCs: la creciente resistencia a los tratados de libre comercio y los acuerdos bilaterales de inversión* http://www.combatiendolostlc.org, Accessed July 18, 2013.

FAO. *El agua y la seguridad alimentaria.* Roma: FAO-UNWATER, 2012.

Friedmann, Harriet. "Distance and durability: Shaky foundations of the World Food Economy". In *The global restructuring of agri-food systems,* edited by Philip McMichael, 258-276. USA: Cornell University Press, 1994.

McMichael, Philip. "Tensions between national and international control of the world food order: contours of a new food regime", *Sociological Perspectives* Studies in the New International Comparative Political Economy, vol. 35, no. 2 (1992): 343-365.

McMichael, Philip. "Market Civilization and the Neo-liberal Food Regime's Global Food Crisis." Paper Presented at the annual meeting of the International Studies Association 50th Annual Convention "Exploring The Past, Anticipating The Future", New York, February 15-19, 2009.

Mekonnen, Mesfin and Arjen Hoekstra. "The green, blue and grey water footprint of crops and derived crop products". In *Value of Water Research Report Series,* no. 47. Delf, The Netherlands: UNESCO-IHE, 2010.

Puyana Alicia and José Romero. *Diez años con el TLCAN. Las experiencias del sector agropecuario mexicano*. México: El Colegio de México/FLACSO, 2008.

Renault, Daniel. "Value of virtual water in food: principles and virtues." Paper presented at the UNESCO-IHE Workshop on Virtual Water Trade, FAO. The Netherlands, December 12-13, 2002.

Schaeffer, Robert. "Free trade Agreements: their impact on agriculture and the environment". In *Food and Agrarian Order in the World-Economy* edited by Philip McMichael, 255-274. USA: Greenwood Press, 1995.

Secretariat of Economy. *Information on foreign trade and foreign direct investment,* http://www.economia.gob.mx/comunidad-negocios/comercio-exterior, Accessed August 5, 2013; http://www.economia.gob.mx/comunidad-negocios/inversion-extranjera-directa, Accessed August 5, 2013.

Warner, Jeroen. "Virtual water -virtual benefits? Scarcity, distribution, security and conflict reconsidered". In *Virtual Water Trade, Proceedings of the international expert meeting on Virtual Water Trade*, edited by Arjen Hoekstra, Value of Water Research Report, Series 12, 125-135, The Netherlands: UNESCO-IHE, 2003.

Part IV
Applying the WF Approach for Impact Analysis on Sectors and Regions

Chapter 17
Water Footprint of Four Cereals in Irrigation District 011

Rosario H. Pérez-Espejo and Thalia Hernández-Amezcua

Abstract In this chapter, the water footprints (WFs) of corn, wheat, sorghum, and barley are estimated in Irrigation District 011 Alto Rio Lerma (DR011), Guanajuato, Mexico, cereals that cover 90 % of the cultivated area, a little more of the water extracted, and generate 70 % of the production value of DR011. The DRiego and Cropwat Programs were used and their results are compared with those obtained by Hoekstra for Guanajuato. The results show that the WF estimates are very sensitive to performance and that a low WF does not indicate the actual water use. Wheat WF is virtually identical for Cropwat and Hoeskstra and it differs by less than 10 % with that obtained with DRiego. It is considered that estimates of barley WF obtained from DRiego are more logical, because in Guanajuato it is planted mostly using irrigation and with greater yields than rainfed. For the vast differences between the estimates of maize and sorghum obtained with both programs and comparing them with Hoekstra, there is no convincing explanation. These differences are even greater when considering blue and green WFs.

Keywords Water footprint · Cereals · Irrigated agriculture

17.1 Characteristics and Importance of Irrigation District 011 Alto Rio Lerma

Irrigation District 011, Alto Rio Lerma (DR011) is in the Lerma–Chapala Basin, VIII Hydrological-administrative region, Lerma–Santiago–Pacific (or Lerma–Chapala–Pacific), which is one of the largest in the country (2 % of the country); it has 15 million inhabitants (16 % of the national population); it has a significant economic weight (contributes 47 % of census gross value added) and the uses, quality, and supply of water constitute its main problem (Conagua 2009; Cotler et al. 2006; SEMARNAT 2001).

The Lerma River is the largest river supplying the Lerma–Santiago–Pacific basin (Fig. 17.1); this hydrological system provides important environmental services to the basin and it is the source that feeds five of the seven largest lakes in the country:

© The Author(s) 2016
R.H. Pérez-Espejo et al. (eds.), *Water, Food and Welfare*,
SpringerBriefs in Environment, Security, Development and Peace 23,
DOI 10.1007/978-3-319-28824-6_17

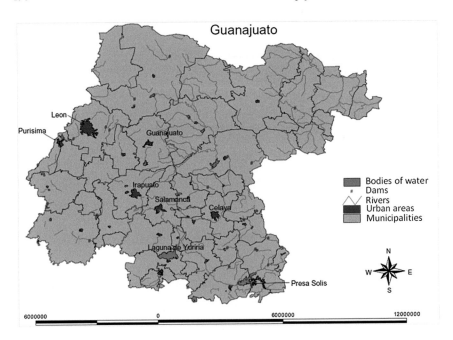

Fig. 17.1 Hydrological system of the Lerma–Santiago–Pacific. *Source* Conagua (2007)

Chapala, Patzcuaro, Cuitzeo, Yuriria lagoon, and Nabor Carrillo (SEMARNAT 2006).

The Lerma subbasin in the states of Mexico, Guanajuato, Michoacan, and Jalisco, is the second most polluted in the country, after the subbasin of the Atoyac River in the states of Tlaxcala and Puebla (Conagua 2009). Agriculture, the main water user, along with the refinery of Petroleos Mexicanos (Pemex) and the electricity plant of the Federal Electricity Commission (CFE in Spanish), both in the city of Salamanca, as well as numerous industries that do not comply with regulations on wastewater discharges, are the most important sources of water pollution.

DR011 located in Guanajuato (Fig. 17.2) is one of the most important agricultural states, which dedicates just over 3 million hectares to agriculture and livestock representing 59 % of its territory (34 % to agriculture and 25 % to livestock). Guanajuato is a small state representing only 2 % of the national territory; however, it ranks ninth in agricultural production value and fifth in livestock (SAGARPA 2011). Primary activities represent 7 % of its gross domestic product (GDP) and place Guanajuato among the 10 states with the highest primary GDP, after large entities such as Jalisco, Sinaloa, Veracruz, Michoacan, Mexico, Sonora, Chiapas, and Chihuahua.

Of the 85 Irrigation Districts of the country, DR011 ranks sixth in extension and is the most important of the Lerma–Santiago–Pacific basin and of the state of Guanajuato (Table 17.1). 87 % of the water extracted in this state is dedicated to agriculture; this situation, coupled with the growing demand of urban and industrial

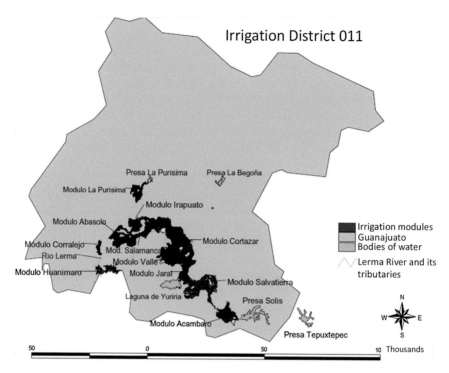

Fig. 17.2 Guanajuato and irrigation district 011 Alto Rio Lerma. *Source* Conagua (2007)

Table 17.1 Top ten irrigation districts, 2008

Code	Name	Hydrological-administrative region	State	Total surface area (ha)
025	Lower Rio Grande	VI Rio Bravo	Tamaulipas	248,001
075	Fuerte River	III Northern Pacific	Sinaloa	227,518
010	Culiacan-Humaya	III Northern Pacific	Sinaloa	212,141
014	Colorado River	I Baja California Peninsula	Baja California and Sonora	208,805
017	Lagoon Region	VII Central Basins of the North	Coahuila de Zaragoza and Durango	116,577
011	Upper Lerma River	VIII Lerma–Santiago–Pacific	Guanajuato	112,772
063	Guasave	III Northern Pacific	Sinaloa	100,125
026	Lower San Juan River	VI Rio Bravo	Tamaulipas	86,102
005	Delicias	VI Rio Bravo	Chihuahua	82,324
097	Lazaro Cardenas	IV Balsas	Michoacan de Ocampo	71,593

Source http://www.conagua.gob.mx/atlas/usosdelagua32.html

sectors, has resulted in enormous pressure on water resources and it has caused the average annual availability per capita to fall from 2800 l in 1950 to 1500 in 1975 and just over 800 at the end of the 1990s. The number of wells exploited, most unauthorized, increased from 2000 in the early 1950s to 16,000 in the late 1990s, and the groundwater level descends more than two meters per year on average (Sandoval-Minero/Almeida-Jara 2006).

Irrigation is a key factor in agriculture in Guanajuato. In 2007, just over one million hectares (Mha) were planted, and of these hectares about 500,000 (48 %) had irrigation surface with rank 3, after Sinaloa (8.1 Mha) and Sonora (5.1 Mha). Half of the cultivated area in Guanajuato provides 85 % of the agricultural production value and DR011 generates much of this value.

The decreasing water availability per capita, its pollution, overexploitation of wells, and the pressure exerted on the resource are factors that make the Lerma–Santiago–Pacific basin, in Guanajuato and DR011 in particular, areas of high vulnerability that threaten the health of people and the ecosystem, and they can be a powerful brake for the future economic activity in the region.

Guanajuato has two Irrigation Districts, 85 La Begona, in the northern part with 15,000 ha of irrigation and DR011 with 115,000 ha that are planted in two agricultural cycles; it has about 22,000 users organized in 11 irrigation modules, 10 of them are located in Guanajuato and one in Michoacan (Conagua 2009). DR011 is one of the oldest in the country, one of the most studied, among other reasons, because it presents a substantive condition of water deficit (Cruz et al. 2002; Kloezen et al. 1997; Vargas 2010).

In the agricultural year 2007–2008, DR011 received 4 % of the total volume of water distributed in the country, i.e., just over one billion cubic meters coming from storage dams (55 %) and groundwater (35 %), a situation that makes this district different from others, where water of wells for irrigation does not exceed 10 %; the remaining water is pumped directly from currents. The official value for gross irrigation depth (irrigation depth extracted from the supply source) for the DR011 was 115 cm, less than 5 cm of the national average irrigation depth (Santos 2012).

DR011 has about 450 km of primary channels and 1192 km of secondary channels; Antonio Coria is the most important channel, with 118 km.

Mexico is a major exporter of vegetables. In 2007, it ranked fourth by the value of its exports and first in the Americas. 89 % of vegetable production is done with irrigation (Financiera Rural 2008: 4), which means a significant transfer of virtual water that is extracted from the northwestern region of the country and from the Bajio, mainly from DR011, regions with a major water deficit.

However, despite the importance of vegetables in DR011, in this first approach to the estimation of the water footprint in irrigated areas, the WFs of only four grains will be analyzed: corn, sorghum, wheat, and barley, because their production occupies 90 % of the cultivated area, an almost equal proportion of water and 70 % of the production value (Santos 2012).

17.2 Estimated Water Footprint of Four Cereals

In this section, irrigation depths and evapotranspiration (ETc) obtained with CropWat (FAO) and DRiego (INIFAP-CENID-RASPA, Technical Bulletin no. 7, February 2007) are compared, to estimate and compare water footprints of maize, sorghum, wheat, and barley in DR011, Guanajuato, and to contrast them with estimates of Hoekstra (2004) for that state.

17.2.1 Methodology

The DRiego Program provides basic information on water requirements and irrigation schedule of annual and perennial crops in the 85 Irrigation Districts (IR) of Mexico. The algorithm uses a method based on the maximum and minimum temperatures, precipitation, and solar radiation; irrigation schedule is the result of a balance of water in the soil and the average climatic information for the past 20 years in each Irrigation District.

The general information required by the program are: federal state, irrigation district, and cultivation. Based on previous elections, the program loads a planting date and the number of days of the growing season (harvest–sow). The program requires information on soil type, moisture constants, content of sand, clay and organic matter, and soil texture (clayey, silty, sandy, loam). Information about cultivation stages and initial available moisture are loaded automatically by the program based on previous elections.

17.2.2 Results

The DRiego Program provides the following information: irrigation schedule; number and date of risks and timeframe between them; irrigation depth and cumulative irrigation depth; daily water balance; maximum and minimum temperature; precipitation; maximum ET, maximum accumulated ET, actual ET, actual cumulative ET; effective precipitation and available soil moisture (DHA); ET charts and accumulated precipitation; variation charts of available soil moisture.

The CropWat Program calculates water and irrigation requirements of crops based on climatic variables, soil characteristics, and crop data. It can be used to estimate water requirements of crops under irrigated and rainfed conditions.

17.2.3 Information Used and Its Sources

The information used by the two programs was:

1. Climate and initial evapotranspiration (ETo):

    i. Minimum and maximum temperature, moisture, and wind: climatic data from the Agroclimatological Station Network of the states of the National Institute of Forestry, Agricultural and Livestock Reseach (INIFAP in Spanish), averages for 2006–2010. The stations consulted were: Acambaro (San Lorenzo); Celaya (INIFAP-CEBAJ); Cortazar (Villa de Cortazar); Salamanca (Los Aguijares); Salvatierra (Huatzindeo); Valle de Santiago (Villadiego); Yuriria (San Vicente);
    ii. Sun exposure: official information only provides a datum, which is the average of 1980–2000 of the National Weather Service (SMN in Spanish, Conagua, Mexico);
    iii. Precipitation: Agroclimatic Stations of the states of the National Institute of Forestry, Agricultural and Livestock Research.
    iv. Crop coefficients (Kc): FAO, for each crop with the following values: corn: 0.3, 1.2, 0.35; sorghum: 0.3, 1.00, 0.55; wheat: 0.3, 1.15, 0.3; barley: 0.3, 1.15, 0.25;
    v. Soil: the information of FAO for medium soils was used.

2. Planting dates (PD): DRiego Program for each of the crops: corn Spring–Summer (S–S) cycle, 01/03; sorghum S–S cycle: 01/03; wheat Fall–Winter (F–W) cycle: 15/12; barley F–W cycle: 15/12;
3. Stages of the growing season (harvest–sow): DRiego Program: corn: 130 days; sorghum: 120 days; wheat: 140 days; barley: 110 days;
4. Irrigation schedule: DRiego Program, moment: critical exhaustion irrigation/No irrigation; application: reset at field capacity; field efficiency: 70 %
5. Finally, crop water requirements: evapotranspiration (ETc); effective precipitation, and irrigation requirements, the latter obtained from DRiego;
    The following additional information was needed to estimate WF:
6. Yield: *Anuario agropecuario* (Agricultural and Livestock Yearbook) SIAP-SAGARPA (1980–2010). Average for the period 2006–2010.
7. Harvested area: *Anuario agropecuario* (Agricultural and Livestock Yearbook) SIAP-SAGARPA (1980–2010). Average for the period 2006–2010.
8. Grain type: *Anuario agropecuario* (Agricultural and Livestock Yearbook) SIAP-SAGARPA (1980–2010). Irrigated and rainfed white grain corn in; irrigated and rainfed grain sorghum; irrigated soft grain wheat; irrigated grain barley (Tables 17.2, 17.3 and 17.4).

17.2.4   Results of ET and WF for DR011

Table 17.2 Water footprint estimates

CropWat	Corn		Sorghum		Wheat	Barley
	Irrigated	Rainfed	Irrigated	Rainfed		
Gross irrigation depth	550.74	–	445.07	–	423.49	556.81
Total net irrigation depth	253.33	–	219.03	–	194.8	256.14
Actual water use	465.3	330.3	410.36	378.84	439.83	454.16
Potential use of water	465.3	465.31	410.36	410.36	439.83	454.16
Yield	8.51	1.95	8.39	2.82	6.25	2.5
Area	4761.10	5419.72	7112.20	2663.37	3529.56	4426.75
BET	253.33	–	219.03	–	194.8	256.14
GET	211.97	330.3	191.33	378.84	245.03	198.01
BWF (m3/ton)	297.53	–	261.15	–	311.82	1024.57
GWF (m3/ton)	248.96	1693.85	228.12	1342.05	392.23	792.06

Cropwat Program. DR011. *Source* The authors
BET Blue evapotranspiration; *GET* Green evapotranspiration; *BWF* blue water footprint; *GWF* green water footprint

Table 17.3 Water footprint estimates

DRiego	Corn	Sorghum	Wheat	Barley
Gross irrigation depth	1243.48	1195.65	969.78	717.39
Total net irrigation depth	572	550	446.1	330
Actual water use	641	599.6	486.5	359.5
Potential use of water	641	599.6	486.5	359.5
Yield	8.51	8.39	6.25	5.19
Area	4761.10	7112.20	3529.56	4426.75
BET	572	550	446.1	330
GET	69	49.6	40.4	29.5
BWF (m3/ton)	672.15	655.77	714.09	636.01
GWF (m3/ton)	81.08	59.14	64.67	56.86

DRiego Program. DR011. *Source* The authors
BET Blue evapotranspiration; *GET* Green evapotranspiration; *BWF* blue water footprint; *GWF* green water footprint

Table 17.4 Water footprints

Corn

WF (m³/ton)	CropWat	DRiego	CropWat rainfed	Hoekstra
Green	248.96	81.08	1693.85	1874.00
Blue	297.53	672.15	–	87.00
Total	546.49	753.23	1693.85	1961.00

Sorghum

WF (m³/ton)	CropWat	DRiego	CropWat rainfed	Hoekstra
Green	228.12	59.14	1342.05	1562.00
Blue	261.15	655.77	–	69.00
Total	489.27	714.9	1342.05	1562.00

Wheat

WF (m³/ton)	CropWat	DRiego	Hoekstra	
Green	392.23	64.67	392.23	
Blue	311.82	714.09	311.82	
Total	704.05	778.76	704.05	

Barley

WF (m³/ton)	CropWat	DRiego	Hoekstra	
Green	792.06	56.86	790.00	
Blue	1024.57	636.01	900.00	
Total	1816.63	692.87	1690.00	

Corn, sorghum, wheat, and barley. DR011. Comparison between CropWat, DRiego and Hoekstra.
Source The authors

17.3 Conclusions

This first approach to the estimation of blue and green water footprints of cereals in an irrigation zone of Mexico provides the following conclusions:

(1) WF, an indicator that relates the amount of water used (m³) with the volume produced (ton) of a crop, is highly sensitive to the yield variable. It is possible to change the magnitude of some data required by the Cropwat and DRiego Programs, and the results hardly change, but any modification in the value of yields further alters the estimated WF.

(2) Low WF does not indicate the actual water use; it leaves out the effects on its quality (overuse of water and agrochemicals in irrigated areas) and vulnerability of aquifers.

(3) WF estimated for wheat using Cropwat is almost identical to those obtained by Hoekstra for the state of Guanajuato and they differ by less than 10 % of those calculated with the DRiego Program. It can be assumed that this is because wheat is a crop more internationally standardized and the technological package used is virtually the same worldwide.

(4) The total WF estimated for barley using DRiego is 41 % lower than that obtained by Hoekstra and 61 % lower than that estimated by Cropwat. The blue and green WFs estimated with DRiego are quite different from those of Cropwat and Hoekstra. However, intuitively, it is considered that a green and total WF estimated with DRiego are much lower than those obtained by Hoekstra and Cropwat, also they are more logical because 80 % of barley grown in Guanajuato is irrigated and yields are six times greater than rainfed. It is assumed that WF estimated by Hoekstra and Cropwat are biased in relation to the low yields in rainfed areas.

(5) For the vast differences between maize and sorghum estimates done by both programs and comparing them with Hoekstra, there is no convincing explanation. These differences are even greater when comparing blue and green WF.

(6) It is not possible to compare rainfed maize and sorghum using DRiego; therefore, the WFs of maize and sorghum obtained by Cropwat and Hoekstra are compared, and of irrigation between DRiego and Cropwat are also compared. For rainfed maize and sorghum, Hoekstra estimates are greater than those of Cropwat by about 16 %. For irrigated maize and sorghum, DRiego estimates are greater than those of Cropwat by 37–46 percent.

(7) For maize and sorghum, blue WF of DRiego is greater than the estimated with Cropwat and the opposite happens with green WF.

(8) Getting different results may lead to an erroneous estimation of final consumer products that are made from these cereals: tortillas, bread, and other foods based on wheat flour, beer, animal products in the case of sorghum.

(9) Finally, there is no accurate information of how Hoekstra estimated the WF of these cereals, and it must be stressed that his estimates are for the entire state of Guanajuato, while those of DRiego are only for Irrigation District 011, from where most of the production of maize, sorghum, wheat, and barley is obtained.

References

* indicates internet link (URL) has not been working any longer on 8 February 2016.

Catalán, Ernesto. *Programa para la calendarización del riego parcelario.* Gómez Palacio, Durango, México: CENID RASPA INIFAP, 2002.
CONAGUA. *Distritos de riego de la República Mexicana.* México: Sistema Integrado de Gestión Administrativa, 2007.
CONAGUA. *Estadísticas agrícolas de los distritos de riego, Crop Year 2007/08,* México: 2009.
Cotler, Helena, Marisa Mazari and Jose de Anda. *Atlas de la cuenca Lerma-Chapala. Construyendo una visión conjunta.* México: SEMARNAT/INE/UNAM/Institute of Ecology, 2006.
Cruz, Valentín, Ramón Valdivia and Christopher Scott. "Productividad del agua en el DR011, Alto Río Lerma." *Agrociencia* (2002): 483–439.
*Financiera Rural. "La producción de hortalizas en México." http://www.financierarural.gob.mx/informacionsectorrural/Documents/Hortalizas.pdf, Accessed July 26, 2012.

Kloezen, Wim, Carlos Garcés-Restrepo and Sam Johnson III. "Los impactos de la transferencia del manejo del riego en el Distrito de Riego Alto Río Lerma, México" *Research reports* 15, Mexico, International Water Management Institute (1997): 5–34. http://publications.iwmi.org/pdf/H_22446.pdf, Accessed July 7, 2012.

Sandoval, Ricardo and Raúl Almeida. "Public policies for urban wastewater treatment in Guanajuato, Mexico." In *Water quality management in the Americas,* edited by Cecilia Tortajada, Asit Biswas, Benedito Braga, Diego Rodríguez, 147–166. México: Springer, National Water Agency, 2006.

Santos, Andrea. "Efectos de la apertura comercial de la economía mexicana en el consumo de alimentos en los hogares urbano-populares, 1992-2010", MA thesis in sociology diss., FLACSO, 2012.

SAGARPA. *Programa Nacional de Medio Ambiente y Recursos Naturales 2001-2006.* México: Semarnat, 2006.

SAGARPA. "Acuerdo por el que se da a conocer el estudio técnico de los recursos hídricos del área geográfica Lerma-Chapala." In *Official Government Gazette,* 8–50. México: Semarnat. http://www.CONAGUA.gob.mx/CONAGUA07/Noticias/EstTecLermaChapalar.pdf, Accessed October 13, 2011.

Vargas, Sergio. "Aspectos socioeconómicos de la agricultura de riego en la cuenca Lerma-Chapala." In *Economía, Sociedad y Territorio*, 231–263. Mexico: El Colegio Mexiquense, 2010.

Chapter 18
Forage Water Footprint in the Comarca Lagunera

Ignacio Sánchez-Cohen, Gerardo Delgado-Ramírez,
Gerardo Esquivel-Arriaga, Pamela Bueno-Hurtado
and Abel Román-López

Abstract This chapter presents the bovine milk production status in the Comarca Lagunera, arid region of north–central Mexico, which has had a major agricultural/livestock and industrial development, being livestock the activity that stands out. In 2011, 2,274,797 l of milk were produced, which represented 21.2 % of the national production. This production is developed intensively, in an area characterized by shortages and water quality. The objective of the chapter is to determine bovine milk water footprint and the pressure of this production on water resources in the region.

Keywords Water footprint · Forages · Semiarid Mexico

18.1 Status of the National Bovine Milk Production

Milk production in Mexico is very heterogeneous in terms of the technology and agro-ecological and socio-economic conditions in which it develops. The variability of weather conditions acquires regional characteristics that qualify the tradition and customs of the population (Sagar 1999). Milk production in Mexico is obtained from three systems (Mariscal et al. 2004).

1. Intensive dairy. It participates in 54 % of the national production. It comprises companies that produce with high unit costs that require large production volumes and high prices for profit. They use highly productive cattle, mainly Holstein, and they have specialized facilities and mechanized processes.
2. Family dairy. It contributes 31 % to national production. In general, it constitutes an important source of raw material for the dairy industry; however, it is seasonal and temporary for the pasteurizer industry. The industry benefits from a low price and continuous supply; it is a system that softens price increases in times of growth; it has low operating costs and little dependence on external inputs to the company.

© The Author(s) 2016
R.H. Pérez-Espejo et al. (eds.), *Water, Food and Welfare*,
SpringerBriefs in Environment, Security, Development and Peace 23,
DOI 10.1007/978-3-319-28824-6_18

3. Dual purpose dairy. It participates in 15 % of the national milk production. It
   develops mainly in tropical regions of the country using Zebu breeds and their
   cross-breeding with Swiss, Holstein, and Simmental. It is characterized because
   livestock can have two zootechnical functions, to produce meat or milk,
   depending on market prices. The handling is extensive, basing its feeding on
   induced pastures and improved in lesser degree.

From 1980 to 2011, production had a positive annual growth with a slight drop
from 1986 to 1989 (Fig. 18.1).

Based on the estimated production in 2011 (Fig. 18.2), production is concen-
trated in the following milk producing areas: Comarca Lagunera (Coahuila,
Durango) 21.1 % of domestic production; Centre (Jalisco, Guanajuato, State of
Mexico, Hidalgo, and Puebla) 37.9 %; Tropic (Veracruz, Chiapas, Tabasco,
Oaxaca, Campeche, and Tamaulipas) 13.5 %; Chihuahua (Municipalities of
Delicias and Cuauhtemoc) 8.7 %, and the rest of the states, 18.8 %.

From 1993 to date, the number of dairy cattle has had annual rates of positive
growth, increasing from 1,632,552 head of cattle in 1993 to 2,382,443 in 2011
(Fig. 18.3). The largest livestock population is located in Jalisco (13.4 %),
Chihuahua (10.5 %), Coahuila de Zaragoza (10.1 %), Durango (10.3 %), Hidalgo
(8.4 %), Guanajuato (7.8 %), Puebla (7.2 %), and the State of Mexico (4.9 %).
These states concentrate 72.5 % of the national inventory of livestock.

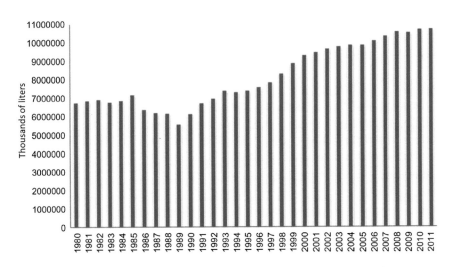

Fig. 18.1 Historical bovine milk production in Mexico. *Source* Authors' information

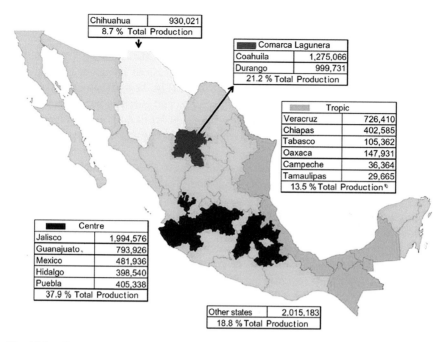

Fig. 18.2 Milk production by region. *Source* SIAP (2011)

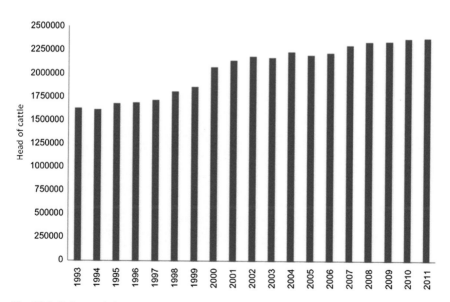

Fig. 18.3 Dairy cattle inventory 1993–2011. *Source* SIAP, SAGARPA (2012)

18.2   Bovine Milk Production in the Comarca Lagunera

The Comarca Lagunera is located in the central part of the northern portion of the country, between southwestern Coahuila and northeastern Durango; it comprises ten municipalities of Durango and five of Coahuila (Fig. 18.4). It is located between the meridians 102°22′ and 104°47′ west longitude, and parallels 24°22′ and 26°23′ north latitude. The average height above sea level is 1,139 m. It has a mountainous extension and a flat surface where agricultural areas are located, as well as urban areas (SAGARPA 2011).

The production system is intensive, with stabled Holstein cattle, fed with cut forage and concentrates. It is supplied with forage produced locally and purchased outside the region. Most producers use artificial insemination and embryo transfer. It has preventive veterinary care and skilled workforce or at least with some experience (Mariscal et al. 2004).

Production requires abundant and good quality forages, which are complemented with concentrated food based on grains; it uses plenty of water to drink and cleaning, but especially for growing forage. Due to housing, a lot of manure is produced whose disposal is costly (Castro et al. 2001).

Regional dairy cattle had grown from 75,092 head of cattle in 1981 to 231,713 in 2010 (Fig. 18.5).

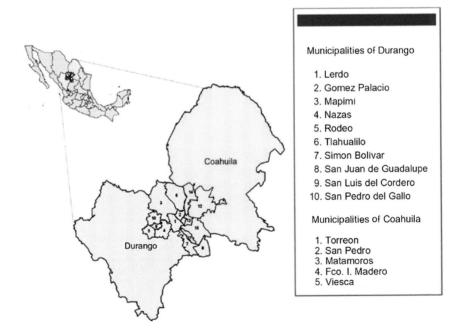

Fig. 18.4 Geographical location of the Comarca Lagunera *Source* Authors' information

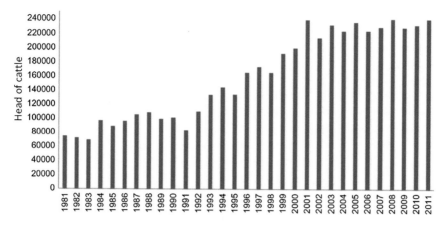

Fig. 18.5 Dairy cattle inventory 1981–2011 Comarca Lagunera. *Source* SIAP, SAGARPA (2011)

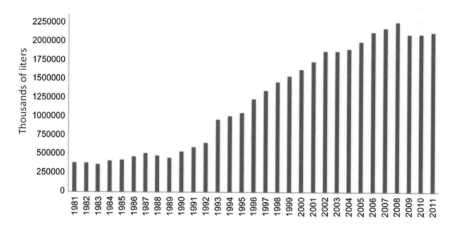

Fig. 18.6 Historical production of bovine milk Comarca Lagunera. *Source* SIAP, SAGARPA (2011)

Milk production in the region has grown steadily since 1981, reaching a peak production of 2,255,273 l in 2008 (Fig. 18.6). The average daily production per cow is 24.1 l.

18.3 Water Footprint of Forage Production in the Comarca Lagunera

Water footprint (WF) calculation relates production to the freshwater volume used to produce goods or products, and it consists of three elements: blue water, green water, and gray water. In arid and semiarid regions, there is greater consumption of blue water for agricultural production (Mekonnen et al. 2010), due to water shortage because of low rainfall. In the Comarca Lagunera, average annual precipitation ranges from 200 to 250 mm and it is necessary to irrigate crops, extracting large volumes of groundwater (blue water) to produce forage, cotton, vegetables, and nuts. Therefore, in this WF estimate, just blue water is calculated; it was necessary to know the diets used in most stables to identify the main forage components and to define an approximate amount of water consumed by cattle to produce one liter of milk.

18.3.1 Dairy Cattle Feeding

Dairy cattle feeding is the factor with more incidence in milk production; a good diet improves milk production, health, and reproduction (Cañas 1998). The amount of food consumed varies according to live weight, production level and lactation time aspects that are considered in the formulation of an optimal diet of forage and concentrate.

The potential performance of a cow is a feature it has since the gestation period and it depends on the quantity and quality of food. Maximum milk production is achieved between 45 and 60 days after giving birth (Fig. 18.7). A cow can potentially produce 25 l at the point of maximum production, but without adequate food it produces only 20 l; this represents 900 l less in total lactation (180 days).

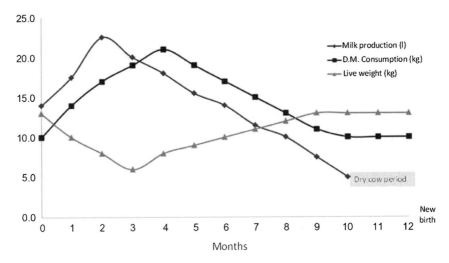

Fig. 18.7 Milk production, consumption, and live weight during lactation. *Source* SIAP, SAGARPA (2011) and authors' information

In general, the maximum food consumption is out of phase with respect to milk production (Fig. 18.7); however, it is possible to manipulate the maximum production point using some type of diet. The first three months represent the most demanding period in cow feeding.

Moreover, at this stage about 45 % of total lactation milk is produced; in the second and third trimesters 32 and 23 % are produced, respectively (Hazard 1990). For this reason, the following dry matter intakes are recommended per kilogram of live weight during different stages of lactation: 1st third = 3.6 % of live weight; 2nd third = 3.0 % of live weight; 3rd third = 2.5 % of live weight.

This means that an animal of 500 kg of live weight should consume, respectively, 18, 15, and 12.5 kg of dry matter during the first, second, and third stages of lactation. Dry matter is the amount of food without the water it contains; it is a unit used to be able to compare the content of different forages and concentrates.

Dairy cattle diets should include water, dry matter, proteins, fibers, energy, vitamins, and minerals in sufficient and well balanced amounts. Livestock water needs depend on age, production, climate, and dry matter intake. Table 18.1 presents water consumption of dairy cattle at different stages; a cow producing 30 l of milk per day consumes the largest amount of water, between 90 and 150 l/day.

Water supply for cattle comes mainly from three sources: water consumed in free form; water ingested in food and water produced by metabolism. On average, it is estimated that 83 % of total water consumed is in free form. Water requirement per liter of milk produced ranges from 2.3 to 3.0 l. In general, cattle must permanently have clean and fresh water, being able to consume between 66 and 115 l/day (Bartaburu 2002).

A bovine animal consumes between 2 and 3 % of its live weight of dry matter, depending on its milk production and usually 2/3 parts of it are given as forage. According to Wattiaux (2002), food is classified as forage, concentrates (energy and protein), vitamins, and minerals. This classification is somewhat arbitrary; it is essential to determine which food is available, its nutritional value, and factors affecting its use in the diet.

Table 18.1 Water requirements of dairy cattle

Animal stage	Water requirements (l/day)
Calves	5–15
Bovine cattle of 1–2 years	15–35
Dry cows	30–60
Cows with milk production of 10 l	50–80
Cows with milk production of 20 l	70–100
Cows with milk production of 30 l	90–150

Source Hazard (1990)

18.3.2   Water Footprint (Blue Water) Estimate in Milk Production

Table 18.2 presents the ingredients that make up three diets; "diet 1" for an average milk production of 21.5 l/day/cow, scheduled for two milkings per day, with a milk fat content of 3.73 %. "Diet 2" is programmed for an average production of 28.5 l/day/cow in two milkings, with a fat content of 3.35 %. Finally, "diet 3" is programmed for an average production of 32.0 l/day/cow in three milkings, with a fat content of 3.27 %.

All three diets are programmed for a dry matter intake greater than 20 kg day^{-1}, which is higher than that recommended by Hazard (1990) of 18 kg day^{-1} in the first third of lactation. Table 18.3 presents the amount of forage required to feed 240,108 cows in the Comarca Lagunera, which corresponds to the inventory of 2011. Values of dry matter for each of the forage ingredients were taken into account to convert

Table 18.2 Common diets for dairy cattle feeding in the Comarca Lagunera

Ingredients	% Dry matter	Diet no. 1 (kg/cow/day)	Diet no. 2 (kg/cow/day)	Diet no. 3 (kg/cow/day)
Alfalfa (hay)	88.0	3.0	5.5	6.65
Alfalfa (green)	25.0		14.0	
Sorghum silage	35.0	24.0	16.0	
Corn silage	35.0			18.5
Rolled corn	85.0	7.68	5.0	7.4
Rolled sorghum	90.0		2.0	
Soy flour	92.0	1.92		
Soy meal	88.0		1.3	1.1
Canola meal	90.0		1.75	
Cottonseed	93.0	0.12	2.0	2.5
Sodium bicarbonate[a]		0.016	0.15	0.2
Energizing salt[a]			0.2	0.4
Minerals[a]			0.12	0.36
Rock salt[a]		0.16		
Magnesium oxide[a]			0.03	
Calcium carbonate[a]			0.03	
Potassium chloride[a]			0.05	
Urea[a]				0.1
Bypass fat[a]				0.36
Total dry matter	20.0	25.0	23.0	

Source Authors information
[a]These ingredients are considered with a 100 % of dry matter

Table 18.3 Forage required for the three diets used in dairy cattle feeding in the Comarca Lagunera

	D.M.[a] (kg/day)	G.M.[b] (kg/day)	R.P.G.M.[c] (t/day)	R.P.G.M.[c] (t/year)
Diet no. 1				
Alfalfa (hay)	2.6	12	2,881	1,051,673
Sorghum silage	8.4	68.6	16,471	6,012,064
Diet no. 2				
Alfalfa (hay)	4.8	22	5,282	1,928,067
Alfalfa (green)	3.5	14	3,362	1,226,952
Sorghum silage	5.6	45.7	10,973	4,005,121
Diet no. 3				
Alfalfa (hay)	5.9	26.6	6,387	2,331,209
Corn silage	6.5	52.9	12,702	4,636,125

Source Authors' information
[a]*D.M.* Dry matter
[b]*G.M.* Green matter
[c]*R.P.G.M.* Required production of green matter

dry matter (DM) weight to green matter (GM) and to estimate the daily and annual required production of green matter (RPGM), under the three different diets.

It is observed that the amount of green forage required is different for each diet, based on the number of milkings per day, animal live weight, daily milk production, and fat percentage it is desired to have in milk, which should range between 3 and 4 % to avoid a penalty in sales price by the pasteurizer company. Also, water consumption of forage crops is different, depending on photosynthetic capacity of each species, climatic conditions, and water availability in the soil.

Estimate of crop water demand depends largely on the knowledge of the amount of water it consumed and the right time to use it, in order to avoid undermining its performance. Alfalfa requires approximately 140 cm (Inzunza 1989); forage corn 54.8 cm (Andrew et al. 2009) and forage sorghum 43 cm (Doorenbos/Kassam 1986). Average yield of these crops in the region is 80, 48, and 50 tons per hectare per year (t/ha/year) (SAGARPA 2011). Alfalfa yield is considered in green forage.

Table 18.4 presents the surface required for forage production (green matter) in each of the diets used and the volume of water required by crops in each diet. It was considered an inventory of 240,108 cows, performance and water consumption of each forage crop.

"Diet 1" would require 11.6 and 6.1 % more of planted surface in relation to "diet 2" and "diet 3", respectively. Regarding water consumption, "diet 1" would use 18.7 and 22.9 % less water compared to "diet 2" and "diet 3", respectively.

In the Comarca Lagunera, around 111,100 ha of irrigated forage are planted: 34 % of alfalfa; 26 % of forage maize, and 24 % of forage sorghum (SAGARPA 2011). With this information and data from Table 18.4 on irrigation surface for forage production, it can be deduced that a portion of water is brought from outside the region (virtual water), in the purchase of forage to meet animal nutritional needs.

Table 18.4 Surface and volume of water required for green forage production in the three diets used for dairy cattle feeding in the Comarca Lagunera

	R.P.A.[a] (ha)	R.W.V.[b] (mm^3)
Diet no. 1		
Alfalfa	13,146	184
Forage sorghum	120,241	577.2
Subtotal	133,387	761.2
Diet no. 2		
Alfalfa	39,438	552.1
Forage sorghum	80,102	384.5
Subtotal	119,540	936.6
Diet no. 3		
Alfalfa	29,140	408
Forage corn	96,586	579.5
Subtotal	125,726	987.5

Source Authors' information
[a]*R.P.A.* Required planted area
[b]*R.W.V.* Required water volume

Table 18.5 presents the blue WF estimate used in milk production in the Comarca Lagunera, based on the following information: diets more common for dairy cattle; maximum expected milk production per cow; number of cows in production in the region (240,108 head of cattle), and water consumption of dairy cattle (freeform and from food).

The estimated BWF to produce a liter of milk in the Lagunera Region ranges from 0.36 to 0.41 cubic meters; this variation depends basically on the diet, being diet three the most efficient in water use. This estimate was made using water consumption required for crops (evapotranspiration) in the Comarca Lagunera, without considering the potential efficiency of the irrigation systems used. The BWF value (0.36–0.41) will increase with irrigation technification. Table 18.6 presents gross irrigation sheet applied in alfalfa, corn and forage sorghum with the irrigation systems most used in the region (surface, sprinkler center pivot type, and drip streak).

Table 18.5 Estimate of blue water footprint on milk production under three diets used in the Comarca Lagunera

	M.E.M.P.[a] (l/day/cow)	E.A.M.P.[b] (Thousands of l and/or t)	T.W.C.[c] (mm^3)	E.B.W.F.[d] (m^3 of water/l of milk)
Diet no. 1	21.5	1,884,247.53	769.53[e]	0.41
Diet no. 2	28.5	2,497,723.47	944.95[e]	0.38
Diet no. 3	32	2,804,461.44	995.80[e]	0.36

Source Authors' information
[a]*M.E.M.P.* Daily maximum expected milk production per cow
[b]*E.A.M.P.* Estimated annual milk production
[c]*T.W.C.* Total water consumption
[d]*E.B.W.F.* Estimate of blue water footprint
[e]This amount also considers free water consumption per cow (95 l day^{-1})

Table 18.6 Gross irrigation sheet applied in major forage crops of the Comarca Lagunera with different irrigation system

Crops	I.S.[a]	G.I.S.A.[b] (m)
Alfalfa	Surface	1.90[c]
	Sprinkler	1.70[c]
	Drip streak	1.30[c]
Forage corn	Surface	1.00[c]
	Sprinkler	0.72[c]
	Drip streak	0.62[c]
Forage sorghum	Surface	0.58[c]

Source Authors' information
[a]*I.S.* Irrigation system used for forage production
[b]*G.I.S.A.* Gross irrigation sheet applied in crops
[c]These irrigation sheets are product of research results validated in the Cenid-Raspa

Table 18.7 Estimate of blue water footprint on milk production under three diets and different irrigation systems used in forage production in the Comarca Lagunera

	I.S.[a]	T.W.C.[b] (mm^3)	E.B.W.F.[c] (m^3 of water/l of milk)
Diet 1	Surface	955.5	0.51
	Sprinkler	929.21	0.49
	Drip streak	876.62	0.47
Diet 2	Surface	1222.24	0.49
	Sprinkler	1143.36	0.46
	Drip streak	985.61	0.39
Diet 3	Surface	1527.85	0.54
	Sprinkler	1199.13	0.43
	Drip	985.98	0.35

Source Authors' information
[a]*I.S.* Irrigation system used for forage production
[b]*T.W.C.* Total water consumption
[c]*E.B.W.F.* Estimate of blue water footprint

Table 18.7 contains WF estimates for each diet, considering irrigation system and with the same information as in the previous estimate: diets, maximum milk production per cow, head of cattle in production, and cattle water consumption.

The WF value is modified according to the diet provided, but it is also different for each irrigation system. In this case, "diet 1" has the best index of WF, using drip irrigation system for alfalfa production and surface irrigation for forage sorghum.

18.3.3 Water Footprint Expressed in Energy Consumption

Irrigation uses well water and an additional problem to the high water consumption for forage production is the consumption of electricity, which in turn requires water for its generation. Hence, an important indicator is the kilowatt hour (kWh) required to produce one liter of milk. This value is a direct function of the electromechanical efficiency of pumps, pumping depth and irrigation sheet applied.

Electrical energy (EE) for forage production with well water represents between 30 and 33 % of production costs (Delgado et al. 2012), so it is important to operate the pumping equipment at maximum power or at least at the minimum efficiency set by the Mexican Official Standard-006-ENER-1995, NOM-006 (Sener 1995), which establishes a minimum efficiency of 64 % for electric motors of 150–350 hp, which are the most used in the Comarca Lagunera.

The wells in the studied region work with an average electromechanical efficiency of 40 %, a significantly low value with respect to the minimum of the NOM-006 (Román et al. 2011). The pumping depth has an average value of 130 m (Romero/Melville 2004), level of extraction that causes high power consumption and high forage production costs. Research on electricity consumption of pumping equipment (INIFAP-CNID-RASPA) generated a mathematical function that determines the energy requirement (kWh) per hectare of irrigation, depending on irrigation sheet and electromechanical efficiency (Román/Sanchez 2004). Table 18.8 shows the EE required in deep wells for different irrigation sheet (Is), considering electromechanical efficiency of equipment (kWh.Is.ha.m). The mathematical function is

$$E.R. = 0.28 \left(\frac{E.E.}{I.S.}\right)^{-0.99} (P.D.)$$

where:

E. R. Electricity required to irrigate one hectare (kWh).
E.E. Electromechanical efficiency of pumping equipment (decimal)
I.S. Irrigation sheet applied to crops (cm)
P.D. Pumping depth (m)

Table 18.9 presents the EE requirement to irrigate one hectare of forage crops (alfalfa, corn and sorghum) with well water. This calculation is based on the gross irrigation sheet applied to the crop in each irrigation event, the number of irrigations throughout the growing season and, especially, the electromechanical efficiency of the pumping equipment.

Sorghum is a culture that requires fewer EE per hectare, even when surface irrigation is used with electromechanical efficiencies from 40 to 64 %. This culture does not meet the nutritional needs of dairy cattle due to the low protein intake; therefore, it must be complemented with alfalfa and/or corn silage. Regarding EE savings by irrigation type, in alfalfa production, 8.7 % of EE can be saved if sprinkler is used and 29 % if surface irrigation is used instead of drip streak. In forage maize, 26.7 % of EE can be saved if gravity irrigation is replaced by a sprinkler system and 35 % if drip streak irrigation is used. These savings do not

Table 18.8 Use of electricity based on electromechanical efficiency of equipment (E.E.) and gross irrigation sheet (IS)

Electromechanical efficiency	Irrigation sheet (Cm)											
	5	10	15	20	25	30	35	40	45	50	55	60
0.05	27.59	55.18	82.77	110.36	137.95	165.54	193.13	220.72	248.31	275.9	303.49	331.08
0.1	13.8	27.59	41.39	55.18	68.98	82.77	96.57	110.36	124.16	137.95	151.75	165.54
0.15	9.2	18.39	27.59	36.79	45.98	55.18	64.38	73.57	82.77	91.97	101.16	110.36
0.2	6.9	13.8	20.69	27.59	34.49	41.39	48.28	55.18	62.08	68.98	75.87	82.77
0.25	5.52	11.04	16.55	22.07	27.59	33.11	38.63	44.14	49.66	55.18	60.7	66.22
0.3	4.6	9.2	13.8	18.39	22.99	27.59	32.19	36.79	41.39	45.98	50.58	55.18
0.35	3.94	7.88	11.82	15.77	19.71	23.65	27.59	31.53	35.47	39.41	43.36	47.3
0.4	3.45	6.9	10.35	13.8	17.24	20.69	24.14	27.59	31.04	34.49	37.94	41.39
0.45	3.07	6.13	9.2	12.26	15.33	18.39	21.46	24.52	27.59	30.66	33.72	36.79
0.5	2.76	5.52	8.28	11.04	13.8	16.55	19.31	22.07	24.83	27.59	30.35	33.11
0.55	2.51	5.02	7.52	10.03	12.54	15.05	17.56	20.07	22.57	25.08	27.59	30.1
0.6	2.3	4.6	6.9	9.2	11.5	13.8	16.09	18.39	20.69	22.99	25.29	27.59
0.65	2.12	4.24	6.37	8.49	10.61	12.73	14.86	16.98	19.1	21.22	23.35	25.47
0.7	1.97	3.94	5.91	7.88	9.85	11.82	13.8	15.77	17.74	19.71	21.68	23.65
0.75	1.84	3.68	5.52	7.36	9.2	11.14	12.88	14.71	16.55	18.39	20.23	22.07

Source Authors' information

Table 18.9 Electrical energy required to irrigate one hectare of the main forage crops of the Comarca Lagunera

Crops	I.S.[a]	G.I.S.A.I.[b] (cm)	N.I.A. C.[c]	E.E.R.I.[d] with 40 % E.E.[e] (kWh/ha)	E.E.R.I.[d] with 64 % E.E.[e] (kWh/ha)
Alfalfa	Surface	19	10	17036.825	10648.016
	Sprinkler	4.5	38	15333.143	9583.214
	Drip streak	0.87	150	11701.609	7313.505
Forage corn	Surface	25	4	8966.75	5604.219
	Sprinkler	5.15	14	6465.027	4040.642
	Drip streak	1.13	55	5572.835	3483.022
Forage sorghum	Surface	19.3	3	5191.748	3244.843

Source Authors' information
[a]*I.S.* Irrigation system used for forage production
[b]*G.I.S.A.I.* Average gross irrigation sheet applied at each irrigation
[c]*N.I.A.C.* Number of irrigations applied throughout the crop growing cycle
[d]*E.E.R.I.* Electrical energy required to irrigate one hectare during the entire growing season. In this case, alfalfa is being considered as an annual crop
[e]*E.E.* Electromechanical efficiency of pumping equipment of deep well
[d]*kWh* kilowatt hour

consider an increase in electromechanical efficiency of 64 %, which would add a 37 % of savings in each irrigation technification system, if the minimum efficiency established by NOM-006 also increases.

Table 18.10 presents the estimated EE requirement to produce one liter of milk, with different diets and irrigation systems. Information from previous estimates is considered, such as diet type, area required for each forage crop to meet diet needs, EE consumption per irrigation hectare of each forage, gross irrigation sheet applied throughout the crop growing cycle, and an average electromechanical efficiency of pumping equipment of 40 %. It also took into account maximum expected daily milk production per cow for each of the diets.

"Diet 3" with drip streak irrigation has the lowest index of EE consumption per liter of milk produced. This index can be reduced 30 %, if electromechanical efficiency increases to the limits of NOM-006. Reduction of irrigation sheets for a greater efficiency in irrigation application is a fundamental part of the improvement process.

It is necessary to rehabilitate, restore, and maintain pumping equipment and/or deep wells; to level agricultural land and to follow the recommendations of adequate irrigation. Alfalfa is the crop that requires greater amount of EE for irrigation of an annual cycle, regardless of the irrigation system used. Therefore, it is considered to be a contaminant production system, based on the EE values required to irrigate one hectare, which range between 11,790 and 16,635 kWh, approximately equivalent to 5.3–7.5 t of carbon dioxide (CO_2) released into the atmosphere, in each irrigation per alfalfa hectare.

The above analysis shows that it is imperative to manage water resources in a sustainable manner, in order to ensure quality and quantity of water for future

Table 18.10 Estimate of electrical energy required to produce one liter of milk, using different diets and irrigation systems for irrigation of forage ingredients

	I.S.[a]	E.E.R.D.[b] (kWy)[e]	E.E.R.P.M[c] (kWh/l of milk)
Diet no. 1	Surface	96,830	0.45
	Sprinkler	94,273	0.44
	Drip streak	88,823	0.41
Diet no. 2	Surface	124,174	0.44
	Sprinkler	116,504	0.41
	Drip streak	110,155	0.35
Diet no. 3	Surface	155,538	0.49
	Sprinkler	122,288	0.38
	Drip	100,370	0.31

Source Authors' information
[a]*I.S.* Irrigation system used for forage production
[b]*E.E.R.D.* Electrical energy required to irrigate forage ingredients of each diet
[c]*E.E.R.P.M.* Estimate of electrical energy required to produce one liter of milk per meter of pumping depth
[d]*E.E.R.P.M.D.* Estimate of electrical energy required to produce one liter of milk by pumping depth (D = 130 m on average in the study region)
[e]*kWy* kilowatt-year = 8,760 kWh

generations, which implies to substantially improve the use and management of water, from forage production (primary product) till the final or commercial product (milk). In addition, it contributes to the development of green agriculture and the reduction of EE consumption and CO_2 emissions.

On the other hand, it is important to provide diets to dairy cattle, forage that use less water and technify agricultural irrigation for its production, which will contribute to a sustainable management of the resource. It should be mentioned that the estimates made on water footprint in milk production only consider the water used in the production of forage and not the water used in milking, pasteurization, and packaging (industrialization) of the product. Some studies mention that water footprint of milk as a final product (consumer) ranges to about 0.94 and 1.0 m^3 of water per liter of milk produced (UNESCO 2006; Quisqueya 2013).

References

* indicates internet link (URL) has not been working any longer on 8 February 2016.

Aldaya, Maité et al. "Importancia del conocimiento de la huella hidrológica para la política española del agua." *Encuentros multidisciplinares* 10 (2008): 8–20.
Arévalo, Diego. *Una mirada a la agricultura de Colombia desde su huella.* Colombia: WWF, 2012.
Bartaburu, Danilo. "La vaca lechera en el verano: sombra, agua y manejo", http://www.veterinaria. org/asociaciones/vet-uy/articulos/artic_bov/nuevos/blank_copia(76)/bov000.htm, Accessed December 18, 2012.

Castro, Luis et al. "Tendencias y oportunidades de desarrollo de la red leche en México." *FIRA Boletín Informativo* 317, vol. XXXIII, ninth period, year XXX, September, México, 2001.

Cañas, Raúl. *Alimentación y nutrición animal.* Santiago de Chile: Pontifical Catholic University of Chile, Agronomy School, Agricultural Collection, 1998.

CAWMA. *Water for Food, Water for Life: A Comprehensive Assessment of Water Management in Agriculture,* London: Earthscan, 2007.

Delgado, Gerardo et al. "Metodología para la evaluación de la eficiencia global del riego en sistemas tipo válvulas alfalferas: caso Región Lagunera." Paper presented at Simposio Nacional "Los recursos agua, suelo y vegetación y su relación con el desarrollo del sector agropecuario y forestal de México", Gomez Palacio, Durango, Mexico, July 2012.

Doorenbos, Jan and Amir Kassam. *Yield response to water. Irrigation and Drainage.* Italy: FAO, 1986.

Godoy, Claudio et al. "Uso de agua, producción de forraje y relaciones hídricas en alfalfa con riego por goteo subsuperficial." *Agrociencia* 37 (2003): 107–115.

Hazard, Sergio. "Sabe usted cómo alimentar sus vacas lecheras." *Investigación y Progreso Agrícola Carillanca* 9 (1990): 38–41.

Inzunza, Marco Antonio. *Requerimientos hídricos de la alfalfa en la fase productiva.* Gómez Palacio, Durango, México: INIFAP-SARH, 1989.

Mariscal, Valentina, et al. *La cadena productiva de bovinos lecheros y el TLCAN.* Guadalajara, Jalisco, México: Universidad Autónoma Chapingo, Department of Animal Science, 2004.

Mekonnen, Mesfin and Arjen Hoekstra. *The Green, Blue and Gray Water Footprint of Crop and Derived Crop Products.* Volume 1: Main Report. Value of water Research Report Series No. 47. The Netherlands: UNESCO-IHE, Institute for Water Education, 2010.

Montemayor, José Alfredo et al. "Producción de maíz forrajero en tres sistemas de irrigación en la Comarca Lagunera de Coahuila y Durango, México", *Agrociencia* 46 (2012): 267–278.

Peter Rogers, Ramon Llamas and Luis Martinez-Cortina. *Foreword of Water Crisis: Myth or Reality?,* ix-x, London: Taylor and Francis Group, 2006.

*Quisqueya. "La huella de agua de siete productos básicos." http://www.quisqueyainternacional.net/de-interes/la-huella-de-agua-de-siete-productos-basicos, Accessed January 3, 2013.

Román, Abel et al. "Modelación del abatimiento de pozos profundos." *Terra Latinoamericana* 29 (2011): 1–10.

Román, Abel and Ignacio Sanchez. *Uso y manejo de bombas de pozo profundo, Instituto Nacional de Investigaciones Forestales, Agrícolas y Pecuarias.* Gómez Palacio, Durango, México: Centro Nacional de Investigación Disciplinaria en relación Agua-Suelo-Planta-Atmósfera, 2004.

Romero Lourdes and Roberto Melville. "Conflicto y negociación por el agua, una mirada sobre el caso Comarca Lagunera", Proceedings of X Congreso Bienal de la Asociación Internacional para el estudio de la propiedad colectiva Oaxaca, México, August 2004.

SAGARPA. "Boletín bimestral de Leche. " vol. VII, no. 4, September-October, México.

SAGARPA. *Anuario estadístico de la producción agropecuaria, Delegation in the Comarca Lagunera (Durango-Coahuila).* Durango, México, 2011.

*SENER *Norma Oficial Mexicana. NOM-006-ENER-1995, Eficiencia energética electromecánica en sistemas de bombeo para pozo profundo en operación.* http://www.sener.gob.mx/res/Acerca_de/nom-006-ener-95.pdf, Accessed January 11, 2013.

SIAP. "Boletín de Leche" January-March, Secretaría de Agricultura, Ganadería, Pesca y Alimentación, México, 2011.

SIAP. "Población Ganadera de México". http://www.siap.gob.mx/index.php?option=com_content&view=article&id=21&Itemid=33, Accessed March 2, 2013.

UNDP. *Human Development Report 2006: Beyond scarcity: Power, poverty and the global water crisis.* New York: United Nations Development Programme, 2006.

UNESCO. "Water Footprints" *UNESCO Water Portal Weekly Newsletter* 145 (2006), http://www.unesco.org/water/news/newsletter/145_es.shtml, Accessed December 28, 2012.

Wattiaux Michael and Terry Howard. *Alimentos para vacas lecheras.* Winsconsin, United States: Babcock Institute for International Dairy Research and Development, University of Wisconsin-Madison, 2002.

Chapter 19
Water Footprint in Livestock

Rosario H. Pérez-Espejo and Thalia Hernández-Amezcua

Abstract In this chapter, the current characteristics of the livestock sector in Mexico, the most important livestock systems, its technification levels, and productive methods are described. The water footprints of beef, pork, poultry meat, egg, and milk are estimated. It is mentioned that Mexico presents the same changes that have taken place globally in relation to livestock production patterns and consumption. It is emphasized that the heterogeneity of livestock systems and the lack of specific and reliable information represent significant obstacles for the estimation of the water footprint of the products analyzed. The different effects of trade openness on the production and consumption of animal products from different species and livestock systems are identified, and the estimates obtained in this chapter are compared with those calculated by Hoekstra for these products.

Keywords Water footprint · Animal production

19.1 Livestock in Mexico

19.1.1 Background

Mexico, like other developing countries, has experienced in the past three decades what the Food and Agriculture Organization (FAO) qualifies as a "livestock revolution," whose characteristics are the universal use of feeding systems and line breeding, the relocation of production processes of the family unit to the large trading company, the use of genetic engineering, and the steady increase in productivity, production, and consumption of products of animal origin (Steinfeld 2002).

The livestock expansion in Mexico took place from the early 1970s to mid-1980s, in a process called "livestockization of agriculture," confirmed by a significant growth of livestock inventories and production of beef and pork, and consumption of livestock products. Exports of beef and live cattle increased, and the area for cattle-raising and forage crop production increased notably (Pérez 1986).

© The Author(s) 2016
R.H. Pérez-Espejo et al. (eds.), *Water, Food and Welfare*,
SpringerBriefs in Environment, Security, Development and Peace 23,
DOI 10.1007/978-3-319-28824-6_19

The economic and financial crisis of the 1980s, the imposition of a new development model of trade openness, and less state participation in the economy, slowed this process. Pastoralism stalled, pig-farming involuted, and only poultry protected by high tariffs and supported politically, maintained a constant dynamism.

19.1.2 Livestock and Free Trade

With trade openness and, in particular, with the entry into force of the North American Free Trade Agreement (NAFTA) in 1994, this change was accentuated; tariff reduction and elimination of trade barriers allowed a greater share of imports in the growing demand for animal products.

The bulk of livestock production for food consists of meat and viscera of bovine porcine, ovine and poultry, milk from bovines and goats, eggs, and honey. There are other species and products, but are marginal; from the 1980s; products from three subsectors (bovine, porcine, and poultry) have represented about 97 % of the value of livestock production (Pérez 2006).

While these three subsectors are predominant, the participation of each one of them in meat production has changed significantly. In the 1982–1984 triennium, pork production accounted for just over half of the total production of the three subsectors; beef, 28.6 % and poultry, 17.6 %. In the 2007–2009 triennium, the weighing of these was 21.3, 30.7 and 47.6 %, respectively (Tables 19.1 and 19.2). The strong protection received by the Mexican poultry, its technification and vertical integration, its high concentration, and homogeneity of its production processes, explain the different trend of this sector in relation to pig-farming, although both sectors have short cycles and they use basically the same food supplies.

Different "positioning' studies of Mexican livestock, among them those of Pérez (1996), recognized livestock as a loser sector in NAFTA negotiations, due to lower competitiveness in the production of food supplies, greater production heterogeneity, and a market structure that values certain products (viscera, tallow, fats, and lard), above the price they have in partner countries. This explains the low rates of growth in livestock production from 1995 to 2009, with the exception of poultry products (Table 19.3).

Table 19.1 Livestock sector structure

	1972–1974	1982–1984	1992–1994	1995–1997	1998–2000	2001–2003	2004–2006	2007–2009
Bovine	732	762	1,289	1,361	1,396	1,472	1,571	1,669
Porcine	645	1,435	838	924	995	1,054	1,092	1,158
Poultry	249	469	1,021	1,330	1,742	2,079	2,417	2,609
Subtotal	1,625	2,666	3,149	3,615	4,132	4,605	5,080	5,436

Three subsectors (thousands of tons)

Source 1972–1990 Pérez E., "Granjas porcinas y medio ambiente. Contaminación del agua en La Piedad, Michoacán, 2006" (Porcine farms and environment. Water pollution in La Piedad, Michoacan, 2006) and 1995 onwards, SIAP, with information from Sagarpa delegations

Table 19.2 Percentage structure of livestock sector

	1972–1974	1982–1984	1992–1994	1995–1997	1998–2000	2001–2003	2004–2006	2007–2009
Bovine	45	28.6	40.9	37.6	33.8	32	30.9	30.7
Porcine	39.7	53.8	26.6	25.6	24.1	22.9	21.5	21.3
Poultry	15.3	17.6	32.4	36.8	42.1	45.1	47.6	48
Subtotal	100	100	100	100	100	100	100	100

Three subsectors. *Source* 1995 onwards, SIAP, with information from Sagarpa delegations

Table 19.3 Livestock production (average annual growth rates)

	1995–1997/1998–2000 (%)	1998–2000/2001–2003 (%)	2001–2003/2004–2006 (%)	2004–2006/2007–2009 (%)
Meat carcass				
Bovine	2.5	5.2	6.3	5.8
Porcine	7.2	5.6	3.4	5.7
Ovine	5.4	18.9	15.8	9.9
Goat	4.9	7.1	3.1	1.6
Poultry	22.6	16.3	14.2	7.5
Turkey	100.0	10.0	−10.1	−4.7
Subtotal	12.4	10.3	9.4	6.5
Milk				
Bovine	13.9	8.3	3.0	5.3
Goat	1.9	11.0	10.4	1.7
Subtotal	13.7	8.4	3.2	5.2
Other products				
Table egg	22.1	13.8	10.3	9.6
Honey	10.3	3.1	−7.0	4.5
Wax unrefined	12.5	5.0	−7.9	0.4
Greasy wool	1.6	3.4	1.0	5.0
Subtotal				

Source SIAP, with information from Sagarpa delegations

The important exports of the livestock sector are honey and cattle for fattening in the United States. It is noteworthy that Mexico has been a traditional supplier of live cattle to the United States since the early twentieth century, representing one of the most interesting trade flows globally. Once the NAFTA eliminated the barriers to this exchange in both countries, Mexico has sent annually more than one million live animals, on average, to the northern neighbor.

The dynamism of intensive systems has meant an increasing dependence of animal feeding on imported inputs such as sorghum, which Mexico imports 45 % of its needs, yellow corn and soybeans that are imported almost entirely. The Mexican

Product	Consumption per capita (kg/year)		
	1986–1993	1994–2007	TMCA (%)
Beef	15.4	17.1	11
Pork	10.7	12.1	13
Poultry	10.6	22.5	112
Edible viscera	5.1	5.4	6
Egg	13	17	31
Whole milk	101.3	103.2	2
Butter	0.7	0.6	−14
Cheese	1.6	1.9	19
Serum	8	13.6	70

Table 19.4 Consumption per capita of livestock products

Source Santos (2002)

balanced food industry ranks sixth in the world in importance and second in Latin America (Sagarpa, s.a.).

If two periods are considered, one pre-NAFTA from 1986 to 1993 and another from 1994 to 2007, it can be seen that consumption per capita of livestock products, except poultry and egg, is not high and it is lower than the averages in developed countries (Table 19.4).

The new consumption structure of livestock products in Mexico is an example of the global tendency to imitate a western agro-nutritional model that only 15 % of the world population consumes, which is not reasonable from an economic and nutritional point of view (Padilla/Le Bihan 1997), and it also causes severe changes in land use and significant environmental impacts (FAO 2006).

19.1.3   Systems and Production Methods in Livestock

Livestock production generates approximately one million permanent jobs and, except for the poultry subsector, the rest of the Mexican livestock is very hetero-geneous, so that highly technified systems coexist with semi-technified and small familiar units (Sagarpa, s.a.).

Livestock area is estimated at 112 million hectares (Mha) of which 47.6 Mha (24 % of the area) are overgrazed; 90 % of pasture and 70 % of bushes have this condition (SEMARNAT 2007). Overexploitation of communal resources, reduction of forest areas and biodiversity, and generation of 20 % of the total methane produced in the country are attributed to the bovine pastoral system (Semarnat 2009).

In the beef cattle industry, two different systems can be distinguished: pastoral, (generally extensive) and feedlot (intensive). The pastoral system has two modes: dual purpose livestock (meat and/or milk production) in dry and wet tropic areas, and calves livestock for export, typical of arid and semiarid states of the northern border. Foods in these systems are basically pastures, poor in the north and much more abundant in the south. Feedlot system, which is also located in some northern states and in the center of the country, bases its food in grains and other agricultural inputs.

The other intensive livestock are poultry farming, part of pig-farming, and part of dairy cattle. Poultry farming is a highly concentrated and specialized sector; three companies produce 52 % of chicken meat and over 35 % of egg (Juárez et al. 2008). The Mexican poultry farming depends entirely on line breeding produced in the United States and it has minimal impact on employment; it is 'dry' because it uses very little water in its production processes and virtually does not have wastewater discharges. Poultry litter,[1] previously subjected to heat treatment, is recycled with bovine and ovine feeding.

In contrast, pig-farming is a heterogeneous sector where large technified units, semi-technified farms, and a large segment of 'backyard' units, very important in the food and income of small farmers, coexist. High concentrations of animals in the technified segment constitute a risk factor for human and animal health, and they have a negative impact on the environment because of the high volumes of wastes that are not managed properly (FAO 2007).

In dairy livestock,[2] three systems are identified: (1) intensive: units with 265 cows per herd on average that produce between 4,000 and 6,000 liters/cow/year. This system generates 54 % of the domestic production and it is located in the northern and central regions; (2) familiar: it contributes 31 % of national production and it is also located in the central and northern areas; (3) dual purpose: it participates with 15 % of national production and it is developed extensively in the dry and humid tropic of the country (Mariscal et al. 2008).

19.2 Water Footprint in Livestock

The heterogeneity of livestock production systems in Mexico makes its use and water consumption to be also heterogeneous. The main limitation to estimate water footprint (WF) of the livestock sector is the lack of official statistics and individual studies that provide sufficient information on production indicators and water use of different livestock systems. The estimates were based on data coming from different sources: official, international obtained in websites, private consultants, producer associations, and specific studies. A description of the production indicators and sources of information is performed for each system.

---

[1]Poultry litter is the excreta of broiler chickens, which is always mixed with bedding material: sawdust, rice husks or soy hulls, ground corncob, etc. (http://teca.fao.org/es/read/4407).

[2]In Chap. 18, the estimation of water footprint of milk production in a particular region of the country, the Comarca Lagunera, is presented.

19.2.1 Bovine

19.2.1.1 Intensive Feedlot System

1. Total inventory: Sagarpa. Intensive system (12 %): *Censo agrícola, ganadero y forestal* (Agricultural, livestock, and forestry census), INEGI (2007).
2. Production indicators: (a) extraction rate: 80 %; (b) weight at feedlot exit: between 440 and 520 kg; (c) average daily food consumption per head (kg/hd/day): 10.7 kg/hd/day; (d) feeding structure[3] (3 % live weight): corn, 40 %; soy, 10 %; molasses, 10 %; forage, 40 % (a ratio of 87.7 % of the WF estimated for corn using DRiego for Irrigation District 011, Guanajuato was taken); others, 0.75 % (WF was not estimated); (e) carcass yield[4]: between 55 and 60 % of live weight; (f) production: Sistema de Información Agropecuaria (Agricultural and Livestock Information System), SIAP-Sagarpa
3. Water consumption: (a) Drink: 50 l/day (see footnote 1); (b) Service: service water is not considered; (c) Dressing[5]: 11 liters per head (l/hd).
4. Water footprint of domestic and imported food supplies.[6]

19.2.1.2 Extensive System

1. Total inventory: Sagarpa. Extensive system (88 %): *Censo agrícola, ganadero y forestal* (Agricultural, livestock, and forestry census), INEGI (2007).

19.2.1.3 Dairy Cattle (Specialized)

1. Inventory: 50.6 % of the total. Source: Dr. Luis Villamar, Secretariat of Agriculture, Livestock, Rural Development, Fisheries and Food, Sagarpa.
2. Production indicators: (a) average daily food consumption per head: 29.400 (k/hd/day); (b) feeding structure: alfalfa: 7.82 kg; silage maize: 78.23 kg; sorghum: 7.14 kg. Source: Dr. José Zorrilla, consultant and professor at the University of Guadalajara and Sagarpa; (c) milk production: between 24 and 30 liters (l) (data of La Laguna, Coahuila); (d) milking days: 270; (e) lifespan: 7 years; (f) average milk production per year: 37,800 l (20 l/hd/day for 270 milking days x 7).

---

[3]Source: Dr. José Zorrilla, professor, University of Guadalajara.
[4]Source: Dr. Jesús Soriano, nutrition consultant.
[5]Source: Alberto Garrido, Polytechnic University of Madrid.
[6]Fuente: Mekonnen and Hoekstra (2010).

3. Water consumption: (a) Drink: 50 l/day; (b) Service: 13.333 l/hd (1,000 l/day for washing milking equipment; 3,000 for room and farmyard hygiene (data for a herd of 300 adult cows). Source: Dr. Eduardo Ganzález, University of Guadalajara.
4. Water footprint of national food supplies. Alfalfa and sorghum estimated with the DRiego Program.
5. Water footprint of imported food inputs: maize and sorghum. Source: Hoekstra, WF in the United States.

19.2.2   Swine

19.2.2.1   Technified System. Source: MVZ. Marco Antonio Barrera-Wagdymar, Sagarpa

1. Inventory. Source: Sagarpa.
2. Production indicators: (a) average daily food consumption per head: 2 kg; (b) feeding structure: sorghum, 60 %; soybean meal, 35 %; others, 5 % (WF was not estimated); (c) production: 43 % of the inventory; 70 % of total production; (d) carcass yield: 70 % (between 80 and 85 K); (e) age at slaughter: between 165 and 170 days; (f) slaughter weight: between 100 and 110 kg.
3. Water consumption: (a) Drink: 4 l/day (2 l x food); (b) Service: 12.5 l/hd/day. Source: Dr. Paul E. Taiganides, private consultant.
4. Water footprint of national food supplies: sorghum estimated with DRiego Program.
5. Water footprint of imported inputs: soy. Source: Hoekstra for the United States.

19.2.2.2   Semi-technified System

1. Inventory: Source: Sagarpa
2. Production parameters: (a) average daily food consumption per head: 2.4 kg/day; (b) feeding structure: sorghum, 65 %; soybean meal, 30 %; others, 5 % (WF was not estimated); (c) production: %; (d) carcass yield: 75 %; (e) age at slaughter: 180 days; (f) slaughter weight: 90 kg.
3. Water consumption: (a) Drink: 4 l/day; (b) Service: 12.5 l/day; (c) Dressing: 450 l/hd.
4. Water footprint of national food supplies: sorghum estimated with DRiego Program.
5. Water footprint of imported inputs: soy. Source: Hoekstra for United States.

19.2.3   Poultry (Intensive Only)

19.2.3.1   Laying Hen

1. Inventory: Source: Sagarpa
2. Production indicators: (a) average daily food consumption per head (kg/hd/day): 110 g/day; (b) feeding structure: sorghum, 59.6 %; soy, 17.5 %; calcium, 8.2 % (without WF); canola, 5 %; soybean meal, 2.5 %; (c) production: first cycle: 80 weeks; second cycle: 110–120 weeks. Total: 195 weeks; (d) average production: 280–360 eggs/year (320 on average).
3. Water consumption: (a) Drink: 0.30 l/day average growth and mature, Source: Canada Plan Service 2001. Assumptions: a nave of 500 square feet (46.45 m^2) uses 60 gallons of water (227 l) Source: Watkins Susan, 2006, Clean Water Lines for flock health; (b) Service: 0.49 l/hd. 10 animals per square meter (UNA).
4. Water footprint of national food supplies: sorghum, own estimate using DRiego Program for Irrigation District 011 Alto Rio Lerma, Guanajuato.
5. Water footprint of imported inputs: soy. Source: Hoekstra for the United States.

19.2.3.2   Broiler Chicken

1. Inventory: Source: Sagarpa
2. Production indicators: (a) average daily food consumption per head (kg/hd/day): 110 g/day; (b) feeding structure: sorghum, 66.1 %; soy, 28.6 %; calcium, 1.3 % (without WF); (c) age at slaughter: 49 days; (d) slaughter weight: 2.5 kg; (e) yield: 1.68 K average; (f) production: Sagarpa.
3. Water consumption: a) Drink: 0.55 l/day (Canada Plan Service, average broiler chicken (rotisserie and market); (b) Service:0.41 l/day; 12 birds per square meter (see point 5).
4. Water footprint of national food supplies: sorghum and maize (46 % nationally); soy: 10 % nationally; protein meal (54 % nationally). Own estimate with DRiego Program for Irrigation District 011, Alto Rio Lerma, Guanajuato.
5. Water footprint of imported inputs: sorghum and maize (54 % imported); soy: 90 % imported; protein meal (45 % imported); canola (100 % imported). Source: Hoekstra for the United States.

19.3   Study Limitations

(1) Private consultants were involved in the study because of the lack of information; their numbers may be correct, but are not official.

(2) The effect of different types of regional indicators of production systems was not considered.
(3) The technified, semi-technified, and specialized classification hides the wide range of production systems in the country.
(4) It was not considered if the water used was surface water or groundwater.

19.4   Conclusions

Water footprints (WFs) estimated in this chapter can hardly be compared with other estimates, for example, with those of Hoekstra, because the methodology used is not known in detail and because he worked for a period of 10 years. In contrast, the WF estimated here are just for one observation, 2010. However, in some livestock, the WFs are similar; in others, the differences are considerable (Table 19.5).

Correspondence with Hoekstra categories:

1. Bovine, live pure-bred breeding, INDUSTRIAL.
2. Bovine, live pure-bred breeding, GRAZING.
3. Eggs, bird, in shell, fresh, preserved, or cooked, INDUSTRIAL.
4. Poultry, live except domestic fowls, weighing more than 185 g, INDUSTRIAL.
5. Swine, live except pure-bred breeding weighing more than 50 kg, INDUSTRIAL.
6. Swine, live except pure-bred breeding weighing more than 50 kg, MIXED.
7. Milk not concentrated and unsweetened not exceeding 1 % fat, MIXED.

In this first approach of livestock WF, estimates for four branches of the livestock sector were made in seven livestock systems: beef bovine intensive and extensive systems; poultry meat and farming laying hens; technified and semi-tecnified pig-farming, and dairy bovine in specialized system.

It is known that the gray WF of livestock is very high due to the highly polluting discharges from intensive livestock; however, there is no available information to make a first approximation.

19.4.1   Green Water Footprint (GWF)

(1) In extensive bovine livestock, farming laying hens, technified pig-farming and production of specialized milk, GWF has small differences with those of Hoekstra, less than 10 %, and in all cases, except in dairy farming, the estimation is 1 % greater.

Table 19.5 Comparison of water footprints: Hoekstra versus estimates of this study

Species	System	Mexico WF (a)			Hoekstra WF (b)			Hoekstra/Study estimate		
		Blue WF	Green WF	Total WF	Blue WF	Green WF	Total WF	Blue Dif.	Green Dif.	Total Dif.
Bovine	Intensive (1)	568.25	7755.18	8323.43	645.25	8131.66	8776.91	113.55	104.85	105.45
	Extensive (2)	101.75	7637.67	7739.42	94.00	8349.00	8443.00	92.39	109.31	109.09
Poultry	Laying hen (egg) (3)	150.62	3140.32	3290.94	300.00	3286.00	3586.00	199.18	104.64	108.97
	Broiler chicken (4)	294.72	4881.62	5176.34	228.00	2657.00	2885.00	77.36	54.43	55.73
Swine	Technified (5)	303.47	4693.66	4997.13	526.18	5046.34	5572.52	173.39	107.51	111.51
	Semi-technified (6)	431.22	6920.20	7351.42	502.48	3720.73	4223.21	116.53	53.77	57.45
	Average	367.35	5806.93	6174.28	514.33	4383.54	4897.87	140.01	75.49	79.33
Milk	Specialized (7)	176.15	1985.97	2162.12	205.23	1966.02	2171.25	116.51	99.00	100.42

Source The authors

19.4.2   *Bovine*

(2) The GWF of intensive livestock of Hoekstra is 37 % lower than that estimated here and the extensive is 9 % greater. The estimates done here are based on the reference evapotranspiration of the pastures in the semiarid zone of Mexico (Garatuza 2012) and on the forage-grain maize relationship for intensive livestock.

19.4.3   *Pig-Farming*

(3) In the semi-technified system, the GWF estimated here is 27 % greater than that estimated by Hoekstra. This difference may be due to the increased consumption of sorghum in this system.

19.4.4   *Poultry Farming*

(4) Poultry meat industry, the estimate done here is 46 % greater than that of Hoekstra. The difference could be due to the imported content of food supplies and to the proportion of these in the diets.

19.4.5   *Blue Water Footprints (BWF)*

(5) The BWF of Hoekstra for all the systems are estimated, except for extensive bovine livestock and broiler chicken, are greater, in various magnitudes, than those calculated here. In general, there is no information on the consumption of drinking and service water in different livestock; the data used are a combination of direct expert consultation with information from the internet.

19.4.6   *Pig-Farming*

(6) The use of water in technified pig-farming is reasonably quantified in Mexico and it is considered that the data used are an acceptable approximation to reality. However, the estimate of Hoekstra is 73 % greater than the estimate of the study. First, water scarcity and the pressure of authority diminished the use of water in pig-farming and this pressure did not exist in the second half of the 1990s. Perhaps a temporal horizon from the twenty-first century will give a better approach to estimate BWF in this sector.

19.4.7 Farming of Laying Hens

(7) The BWF estimate of Hoekstra is twice the estimate of the study. Poultry farming in Mexico is almost dry; poultry litter is not diluted in water as it is done in the systems of other countries, and is used as fertilizer. Service water is minimal.

19.4.8 Poultry Meat Production

(8) The system is also dry; poultry litter is dry, it is given a heat treatment and it is used in bovine feeding. BWF estimate of poultry meat is greater than that of Hoekstra and, therefore, the above argument does not apply.

19.4.9 Dairy Bovine

(9) BWF and GWF calculated here have no substantial differences with those of Hoekstra.

19.4.10 Total WF (TWF)

(10) Poultry meat production, bovine intensive livestock and pig-farming in semi-technified systems have a TWF greater in no more than 11 % to that estimated by Hoekstra.
(11) TWF of farming of laying hens and semi-technified pig-farming have very large differences with those estimated with Hoekstra, for which there is no explanation.
(12) TWF of dairy livestock is the same as that of Hoekstra.

References

* indicates internet link (URL) has not been working any longer on 8 February 2016.

FAO. *Livestock's long shadow*. Rome: FAO/LEAD, 2006
*FAO. www.fao.virtualcentre.org, Accessed September 12, 2007
INEGI. *Censo agrícola, ganadero y forestal, 2007*. México: INEGI, 2007
Juárez, Ángel et al. "Producción de pollo para carne en México (1980–2002). Estudio descriptivo y análisis de la cadena productiva". In *Presente y futuro de los sectores ganadero, forestal y de*

la pesca mexicanos en el contexto del TLCAN, edited by Alicia Puyana, José Romero and José Antonio Ávila, 223-264. México: El Colegio de México and Universidad Autónoma Chapingo, México, 2008

Mariscal, Valentina et al. "La cadena productiva de bovinos lecheros y el TLCAN". In *Presente y futuro de los sectores ganadero, forestal y de la pesca mexicanos en el contexto del TLCAN,* edited by Alicia Puyana, José Romero and José Antonio Ávila, 223–264. México: El Colegio de México and Universidad Autónoma Chapingo, México, 2008

*Mekonnen, Mefin and Ajren Hoekstra. "The green, blue and grey waterfootprint of crops and derived crop products", In *Value of Water Research Report Series,* no. 47. Delf, The Netherlands: UNESCO-IHE, 2010, www.waterfootptint.org/Reports/Report47-WaterFootprint Crops-Vol.pdf, Accessed Septembre 12, 2013

Padilla, Martine and Genevieve Le Bihan. "La dinamique international de la consommation alimentaire. Économies et Sociétés", *Developpment Agro-alimentaire* 9, (1997): 11–25

Pérez, Rosario. *El Tratado de Libre Comercio de América del Norte y la ganadería mexicana,* México: UNAM, Consejo Mexicano de Porcicultura, 1996

*SAGARPA. *Programa Nacional Ganadero 2007–2012,* www.sagarpa.gob. mx/ganaderia/Publicaciones/Lists/Programa%20Nacional% 20Pecuario/Attachments/1/PNP260907.pdf, Accessed, August 20, 2012

*SEMARNAT. *México. Cuarta comunicación nacional ante la Convención marco de las Naciones Unidas sobre el cambio climático,* www.semarnat.gob.mx.InfSitCap3suelos.pdf, Accessed August 20, 2007

Steinfeld, Hnning. "Responding to the increasing global demand for animal products: Implications for livelihood of livestock producers in developing countries". In *Responding to the increasing global demand for animal products,* edited British Society of Animal Science/American Society of Animal Science/Mexican Society of Animal Production 12–15. Mérida, Yucatán, México: United Kingdom, British Society of Animal Science. 2002

*Winchell, Wayne. *Water Requirements for Poultry,* Canada Plan Service, 5604.2001:4, http:// www.agriculture.gov.sk.ca/5603-Leaf, Accessed June 3, 2012

Public Employees and Experts Consulted

González, Eduardo. Research Professor. Personal communication, June 8, 2012

Marco Antonio Barrera. SAGARPA Personal communication, February 9, 2012

Soriano, Jesús. Animal Nutrition Consultant. Personal communication, March 29, 2012

Taiganides, Paul E. International Consultant in treatment systems. Personal communication, June 22, 2012

Yáñez, María Antonieta. Staff, National Union of Poultry Farmers. Personal communication, June 22, 2012

Villamar, Luis. SAGARPA. Personal communication, February 8, 2012

Zorrilla, José. Animal Nutrition Consultant and Professor of the University of Guadalajara. Personal communication, June 6, 2012

Chapter 20
Water Footprint of Bottled Drinks and Food Security

Roberto M. Constantino-Toto and Delia Montero

Abstract This chapter raises the incongruity between the dynamism of the bottled drinks sector and shortages, quality, and water management in Mexico. The study questions the importance of the bottling sector and the consumption pattern of its products, and it approaches the virtual water content of the products of this industry based on Hoekstra (2010) and Garrido (2010), so that their magnitudes can be compared.

Keywords Water footprint · Bottle soft drinks · Food security

20.1 Introduction

In this chapter, the characteristics of the production of bottled beverages in Mexico are explored. The bottling sector (for soft drinks and water) raises important elements that should be incorporated in the reflections on the following issues: (1) water use; (2) the implications of effective and efficient water supply management; (3) the social effects of water management models; and (4) the imposed restrictions by dietary patterns on the use of natural resources.

The bottling sector is an important area of analysis, first, because of the role it plays in the food sector of the country and its growing economic performance. Second, because the coverage limitations of public drinking water for direct consumption and the dubious quality of the supplied flows have stimulated the growth of the bottled water market, placing it among the largest in the world.

The bottling industry analysis is conducted in two parts: first, it puts into perspective its importance as well as the consumption pattern of bottled water. Second, an initial estimate of virtual water content in the products of this industry and its evolution is made based on Hoekstra (2010) and Garrido (2010), so the magnitudes of the water used in its preparation can be compared.

It is important to indicate that although Mexico has made great efforts to generate statistical information on water, which is required to estimate water footprint and virtual water, it is subjected to the restrictions imposed by the lack of compatibility

R.H. Pérez-Espejo et al. (eds.), *Water, Food and Welfare*,
SpringerBriefs in Environment, Security, Development and Peace 23,
DOI 10.1007/978-3-319-28824-6_20

between statistics produced in the country (see Chap. 4 of this report) and the requirements of specific information required by these methodologies.

20.2 Comments on Bottled Drinks Production in Mexico

According to the international report made by Beverage Marketing Corporation (BMC) in April–May 2010, the Mexican bottled water market was ranked as the largest, with an average consumption of 234 l/person/year (l/year), which represented 13 % of the global water sales, higher than Spain, with 119 l/person/year, and the United States, with 110 l/person/year. OXFAM (2012) estimates that the average consumption of soft drinks per capita in 2011 was 163 l/year, well above the United States, with 118 l/year.

Beyond interpreting these facts as the result of an apparently successful business strategy, it is necessary to put into perspective what may be happening in a country that has an average but decreasing water availability and a significant distributional asymmetry, to develop a bottled water and soft drinks market of such an international magnitude.

The Mexican bottling industry has several production lines: water, soft drinks, fruit juices, new generation drinks (e.g., energy drinks), among others. However, due to its production scale, bottled water and soft drinks are the segments that dominate the industry profile.

Figure 20.1 shows the dimensions of the bottling industry in the context of food and beverages production in the country. As can be seen, this industry represents approximately one-fifth of the production of the national food sector and half the production of the drinks and tobacco industry. Its growth of 6 % in the past 5 years is well above the economic growth of the country.

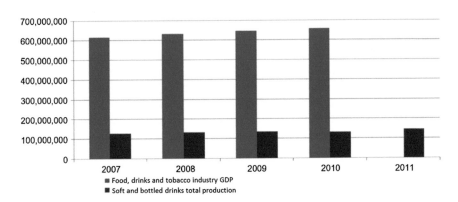

Fig. 20.1 Context of the bottling industry production. Food and bottling industry GDP (thousands of pesos base 2010 = 100). *Source* Based on data from INEGI, National Accounts System 2010

Several factors have contributed significantly to the development of the bottling industry, although there are differences between soft drinks and bottled water. First, and common in both cases, there is an efficient logistics system and distribution of products to the final consumer who is not only present in large cities, but also in communities with less than 2,500 inhabitants. In the particular case of sodas, it stands out that the consumption of these products has become a relatively cheap source of calories and calcium given the asymmetrical distribution of food in the country, Table 20.1 (INSP 2006). Similarly, the unequal distribution of educational capacities in the country contributes to the conformation of this food culture (INEGI 2010).

The national food consumption structure by type of nutrient (Fig. 20.2), as a result of diet changes, is consistent with the changing patterns of morbidity and mortality (Federal Government 2010; INEGI 2012).

A contemporary characteristic of Mexican soft drink market is the inelastic income of its demand (Constantino 2012a); this means that the market has a geographic distribution that tends to behave as the population and not as their income

Table 20.1 Mexico: population by food condition

Access to food	Percentage		Millions of people	
	2008	2010	2008	2010
Food security	53.9	55.7	59.1	62.7
Slight degree of food insecurity	24.4	19.5	26.7	21.9
Moderate degree of food insecurity	12.8	14	14.1	15.8
Severe degree of food insecurity	8.9	10.8	9.8	12.2

Incidence and number of people with food poverty, 2008–2010. *Source* CONEVAL (2010)

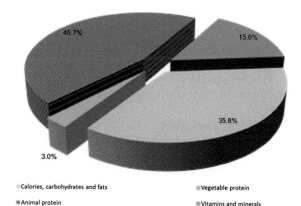

Fig. 20.2 Food expenditure structure by type of nutrient. *Source* Based on data from INEGI. Press Release Num. 270, 2011

45.7%

15.6%

35.8%

3.0%

Calories, carbohydrates and fats   Vegetable protein

Animal protein   Vitamins and minerals

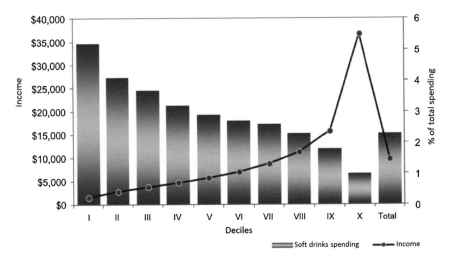

Fig. 20.3 Household spending and income availability. *Source* Based on data from the ENIGH (2004)

level, providing an insight into the household expenditure structure on soda acquisition (Fig. 20.3).

The above feature is important because it determines the solid waste disposal pattern in the territory, in a context of heterogeneous capabilities at local level to manage and facilitate the use and reuse of PET[1] containers, whose volume grows as the soda market size increases (Constantino 2012a).

In the bottled water segment, its precursors in Mexico are closely linked to some water management model characteristics, especially at local levels.

At present, the water sector in Mexico is going through an institutional reconstruction process after a long period of stability (1946–1983), during which a culture with little water caution was generated (Constantino 2012b); at this time, use practices and key management mechanisms were established and rooted, with their corresponding incentive structure.

The lack of financial resources and trained personnel to run the local system operation are the elements that have characterized the institutional transition, from a centralized model to another of deconcentration of public services of drinking water and sanitation at local level (Pineda and Salazar 2008).

This has generated a relative delay in service coverage and a management model that favors the extractive supply over the quality of the supplied flows (Jiménez 2008).

Studies of water quality issues in Mexico (Mazari 2000, 2002, 2005, 2007, 2008; Jiménez 2004) concur that the lack of information and methodological measurement problems prevent to establish a long-term monitoring of surface water and

---

[1]Acronym for Polyethylene Terephthalate.

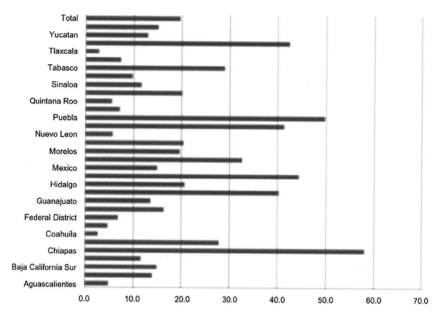

Fig. 20.4 Population at risk due to water quality characteristics. *Source* Based on data from the Federal Commission for Protection against Health Risks

groundwater quality, and they make it difficult to analyze water quality in the distribution points once it has become drinkable or in the networks before delivery to homes. However, it is possible to develop an approach to water quality issues that allows to indicate that it is an issue that requires as much attention as the availability and supply issues.

Jiménez (2008) indicates that there is a tendency of quality reduction in surface water and groundwater, indicated by biochemical oxygen demand (BOD) and chemical oxygen demand (COD).[2] Water quality studies, in which samples from supply sources were taken before and after potabilization (Mazari 2002, 2007), found contaminants that may compromise population health according to the Official Mexican Norm (NOM 127-SSA1-1994), that establishes the quality permissible limits and potabilization treatments for human use and consumption that public and private supply systems must meet.

The potential exposure to health risks arising from water consumption is heterogeneous in the country, as can be seen in Fig. 20.4.

Service coverage deficiencies caused by the limited infrastructure or opacity related to water quality have provided favorable conditions for the emergence and expansion of a bottled water market, not without costly social effects.

---

[2]This approach does not match the information from the National Water Commission.

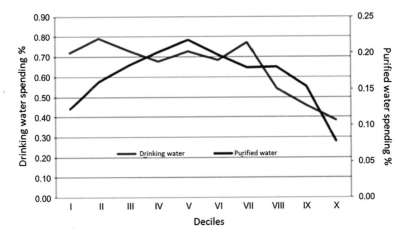

Fig. 20.5 Household spending in water. *Source* Based on data from INEGI. National income and household spending survey, several years

There are two essential features in the presence of a bottled water market in Mexico that can be seen in Fig. 20.5. The first shows that the management model that has allowed its boom contains a regressive strategy in terms of population welfare; the poorest people pay the most expensive water. The second indicates that, along the income distribution, from the poorest to the richest decile, the population is willing to pay more to have water access. This reality questions the diagnosis that indicated that management model inefficiencies were associated to the population reluctance to pay. The incentive structure via price contained in the Public Registry of Water Duties (Repda) indicates that, for all practical purposes, primary sector exemptions constitute a subsidy that inhibits innovation processes and increasing efficiency of water use, particularly in agriculture.

In a scenario as the one above, no price increase for households could compensate the lack of efficiency and stimulus to innovation policy in agriculture related to water.

20.3 The Bottling Industry and Its Relation with Water Footprint and Virtual Water

Until recently, estimates of the amount of water for commercial use were based solely on the quantities used directly in production processes without considering the corresponding sectorial linkages. Water footprint and virtual water consideration has opened a space for alternative reflection that emphasizes the use analysis based on the idea that not all consumed water is equal, and not all used water has a homogeneous social effect (Ercin et al. 2011; Garrido 2010). But more importantly, it shows that there are different ways of combining management tools to improve water availability and supply.

Mexico is not a country with an abundant water supply; however, it does not face an acute shortage. The water that Mexico has would be sufficient to meet public welfare and economic prosperity if the institutional operating conditions were modified. Certainly, the country has public finance constraints that limit the ability to strengthen water sector and its operation. Within the current management model, there is demographic and economic dynamics, sponsored by decentralization incentives, which tend to increase water use competition in northern regions of the country, whose natural availability is lower. In this context, it is useful to explore a sector where water is one of its main direct inputs and, even more important than water, the indirect volume that sustains its dynamics. The resulting lessons could improve the institutional management capabilities.

Social images have been built around the national bottled beverages industry, but especially the soft drinks and bottled water industry that reveal predatory exploitation practices of water resources. Although it is inappropriate to consider business strategies development that put at risk one of its main assets in relation to their performance,[3] the fact is that the lack of public information about the use of a public asset like water in Mexico does not contribute to create trust at social level.

Not all the water used in national bottling industry is groundwater, in the same way that not all water consumed directly in the production process of this industry is the total water used. This is known from the results from studies conducted in the Mexican bottling industry at plant level (González 2007), as well as from international business corporate information.[4]

The water used in the bottling industry production of soft drinks and water in Mexico comes from various supply sources: groundwater, surface water, water public network, or rainwater (González 2007). Given this diversity of supply sources, it is required a tool to quantify freshwater resources appropriation, besides the traditional and restricted "water extraction" measure. Water footprint refers to this total direct and indirect water consumption to make a product (Mekonnen 2011) or alternatively, water footprint network (Coca Cola 2010: 6).

Quantifying only direct water use in production processes without considering the inputs leads to underestimate the magnitude of water resources used in a productive activity and it can hide cross subsidies in manufacturing processes from tax benefits that are granted through water to the primary sector.

---

[3]In 2008, Coca Cola updated the risk for its activities and a requirement for the entire system, which came into force in its bottling plants, was the evaluation of local water resources sustainability used to produce its beverages as well as the sustainability of the available water resources used in the surrounding communities. These assessments include vulnerability, quality and quantity of local water resources (Coca Cola 2010).

[4]Coca Cola uses 153.1 billion liters (km^3) of surface water and groundwater, 139.2 km^3 of municipal water, and 2.2 km^3 of rainwater and other sources to prepare its drinks. The Coca-Cola Company, Sustainability Report Section from 2010/2011.

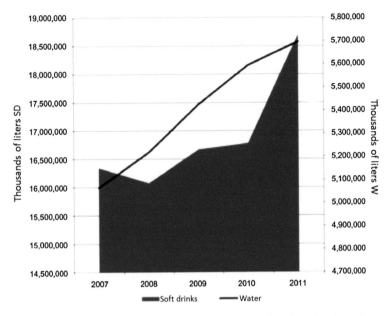

Fig. 20.6 Soft drinks and bottled water sales dynamics. *Source* Based on data from the monthly manufacturing industry survey (several years)

Therefore, in this chapter the Ercin methodology (2011) is applied to measure the water use importance in the bottling industry for water resources and the economy of the country.

Figure 20.6 shows the sales performance of soft drinks and bottled water; it can be observed that this is a dynamic and growing market, and although there are scale differences for each product type, it is clear that most sales of the sector correspond to soft drinks.

When Ercin's technological inputs arrangement (2011) is applied to the Mexican soft drink industry structure, with the variables of water used directly in production and the one used as a high fructose sweetener coming from the United States, it follows that the amount of water used indirectly in the industry is significantly more important than that used directly in the production process (Table 20.2).

Moreover, if the industry strategy related to the bottles utilized in soft drink production is considered, it turns out that water waste caused by the lack of investment to recycle PET waste (post-consumption) is increasing, just as the water footprint of bottle production (Fig. 20.7).

Water footprint calculation offers important information to redesign food strategies and incentives to reduce asymmetries in water use and to mitigate collective welfare distortions.

Table 20.2 Mexican water footprint simulation in soft drink preparation

Inputs for 1 l of soda	Water footprint (l)			
	Green	Blue	Gray	Total
Water input	0	1	0	1
Water in process	0.6	0.6	0	1.2
Water used directly	0.6	1.6	0	2.2
Sweetener (high fructose, USA)	31.8	27.6	13	72.4
Carbon dioxide	0	0.66	0	0.66
Caffeine	105.6	0	0	105.6
Vanilla extract	159.6	0	0	159.6
Lemon oil	0.02	0	0	0.02
Orange oil	1.8	0	0	1.8
Bottle (PET)	0	0.4	8.8	9.2
Screw-on cap (HDPE)	0	0.06	1.36	1.42
Label (PP)	0	0.006	0.136	0.142
Tray transportation	2	0	1	3
Contact film (PS)	0	0.04	0.72	0.76
Flexible wrap for packaging (PS)	0	0.006	0.108	0.114
Packaging labels	0.002	0	0.0008	0.0028
Packing	0.066	0	0.014	0.08
Concrete	0	0	0.01	0.01
Steel	0	0.008	0.1	0.108
Paper	0.0024	0	0.0008	0.0032
Natural gas	0	0	0.048	0.048
Electricity	0	0	0.26	0.26
Vehicles	0	0.002	0.018	0.02
Fuel	0	0	1	1
Water used indirectly	300.8904	28.782	26.5756	356.248
Water Footprint	301.4904	30.382	26.5756	358.448

Technical values of direct and indirect water use in the Mexican soft drink industry. *Sources* Ercin et al. (2011), ANPRAC (2007)

Notes (PET) Polystyrene Terephthalate; (HDPE) High Density Polystyrene; (PP) Polypropylene; (PS) Polystyrene

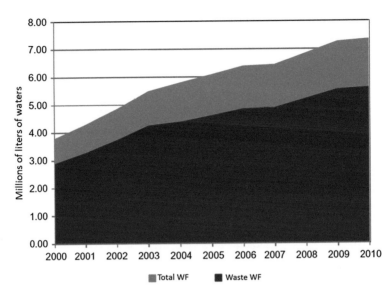

Fig. 20.7 Annual total water footprint of PET packaging and water footprint of its waste. *Source* Based on data from Ercin et al. (2011) and Constantino (2012a)

References

* indicates internet link (URL) has not been working any longer on 8 February 2016.

*Beverage Industry Environmental Roundtable. http://www.anteagroup.com/gcc, Accessed October 4, 2012.

*Beverage Marketing Corporation. http://www.beveragemarketing.com/market-reports.asp, Accessed June 30, 2012.

CONEVAL. *Reporte de medición de la pobreza* en México. México: CONEVAL, 2010.

Constantino, Roberto. *Diagnóstico de la situación actual y análisis de factibilidad de posibles instrumentos económicos para envases de PET posconsumo en México*, México: INE, 2012.

Constantino, Roberto. *Agua y tecnología en México: exploración para un nuevo diseño institucional*, México: UAM-Lerma, 2012.

Ercin, Ertug, Maité Martínez and Arjen Hoekstra. "Corporate Water Footprint Accounting and Impact Assessment: The Case of the Water Footprint of a Sugar-Containing Carbonated Beverage", *Water Resour Manage* (2011):721–741.

Garrido, Alberto et al. *Water footprint and virtual water trade in Spain. Policy Implications.* Natural Resource Management and Policy Series, USA: Fundación Botín, Springer, 2010.

Gobierno Federal. *Acuerdo Nacional para la Salud Alimentaria, Estrategia contra el sobrepeso y la obesidad*, México, 2010.

González-Colín, Mireya, Elena Dominguez and Nydia Suppen Reynaga. "Evaluación técnica, económica y ambiental de la producción más limpia en una empresa de bebidas gaseosas", *Tecnología, Ciencia, Educación* 2, vol. 22, July–December, (2007): 78–83.

Hoekstra, Arjen Ashok Chapagain and Mesfin Mekonnen. *The water foot-print assessment manual: setting the global standard.* London: Earthscan, 2011.

INEGI. *El sector alimentario en México*, México: INEGI, 2010.

INEGI. *Estadísticas de mortalidad*, México: INEGI, 2012.

INEGI. *Encuesta nacional de ingreso y gasto de los hogares*, México: INEGI, 2004–2012.

INEGI. *Sistema de Cuentas Nacionales*, México: INEGI, 2004–2012.

INSP. *Encuesta nacional de salud y nutrición*, México: Instituto Nacional de Salud Pública, 2006.

Jiménez, Blanca. "Calidad del agua en México: principales retos". In *El agua potable en México. Historia reciente, actores, procesos y propuestas*, coordinated by Roberto Olivares and Ricardo Sandoval, 3–28, México: ANEAS, 2008.

Jiménez, Blanca et al. "El agua en el Valle de México". In *El agua vista desde la academia*, 15–32, México: Academia Mexicana de Ciencias, 2004.

Mazari, Marisa et al. "Calidad del agua para uso y consumo humano en la Ciudad de México", XXVIII Congreso interamericano de Ingeniería Sanitaria y Ambiental, mimeo, México: UNAM, 2002.

Mazari, Marisa et al. "Longitudinal study of microbial diversity and seasonality in the Mexico City Metropolitan Area Water Supply System, *Applied and Environmental Microbiology*, 71(9), United Kingdom, 2005.

Mazari, Marisa et al. "Microbiological ground water quality and health indicators in México City", *Urban Ecosystems* 4, (2000)12.

Mazari, Marisa et al. *Impacto de la interrupción del caudal ecológico sobre la calidad del agua*, México: Instituto de Ecología, UNAM, 2007.

Mazari, Marisa and Marcos Mazari. "Efectos ambientales relacionados con la extracción de agua en la megaciudad de México", *Agua Latinoamérica* 2, vol. 8, (2008): 3.

Mekonnen, Mesfin and Arjen Hoekstra. "Mitigating the water footprint of export cut flowers from the Lake Naivasha Basin, Kenya." In *Value of Water Research Report Series*, no. 47, Netherlands: UNESCO-IHE, Delft, 2010.

Mekonnen, Mesfin and Arjen Hoekstra. "The green, blue and grey water footprint of crops and derived crop products." In *Value of Water Research Report Series*, no. 47, Netherlands: UNESCO-IHE, Delft, 2010.

*OXFAM Mexico. http://site.oxfamMexico.org/Mexico-es-ya-el-mayor-consumidor-de-refresco-en-el-mundo-3/, Accessed May 30, 2012.

Pineda, Nicolás and Alejandro Salazar. "De las juntas federales de agua a las empresas de agua: la evolución institucional de los servicios urbanos de agua en México, 1948–2008", In *El agua potable en México. Historia reciente, actores, procesos y propuestas*, coordinated by Roberto Olivares and Ricardo Sandoval, 57–76, México: ANEAS, 2008.

Revista Inversionista (Investor Magazine), No. 297 (2012): 62–64.

The Coca Cola Company, *Product Water Footprint Assessments*, 2010.

The Coca Cola Company, *Sustainability Report Section from the 2010/2011*.

About the Editors

Rosario H. Pérez-Espejo is an economist with an M.Sc. in Agriculture Economics and a Ph.D. in Animal Production and Health from UNAM. She is a full-time researcher with the *Institute of Economic Research* (IER) and member of the National Researchers System Level II. She is professor at both the Graduate School of Economics and the Graduate School of Science in Animal Production and Health at UNAM. She is an expert of the FAO-LEAD (*Livestock, Environment and Development*) Initiative and author of books on economics, international trade, and environmental problems of agriculture in Mexico. She is co-author of books on pig wastewater management; environmental regulations for intensive farms; the economics of nature; agriculture and water pollution. She has published over 110 articles in national and international journals and presented more than 120 papers in Mexico and other countries. Currently, she is developing a project on methane capture and water pollution in intensive pig farms, and on water and food security in Mexico.

Address: Dr. Rosario H. Pérez-Espejo, Circuito Mario de la Cueva, Ciudad de la Investiga-ción en Humanidades, Ciudad Universitaria, C.P. 04510, Mexico, D.F, Mexico.
Email: rosarioperezespejo@gmail.com
Website: www.iiec.unam.mx.

© The Author(s) 2016
R.H. Pérez-Espejo et al. (eds.), *Water, Food and Welfare*,
SpringerBriefs in Environment, Security, Development and Peace 23,
DOI 10.1007/978-3-319-28824-6

Roberto M. Constantino-Toto studied economics at UAM and public policy at the ITAM. He is a member of the Water Network at UAM and of the Scientific Network of Water (Retac) with CONACYT. He works as a researcher on economic policy and natural resource Management at UAM-X. He has been director of the "Raul Anguiano" *Department of Natural Resources and Sustainable Development* (SEMARNAT/INE/UAM) as well as head of the economic policy and development research area at UAM. He has served on several national and international advisory committees regarding water management (CEC/NAFTA, Water System of Mexico City, the Water Sector Scientific Committee, Mexican Institute of Water Technology), and has led international research projects in the field of efficiency in the use of natural resources (REEO-UNEP). His recent publications include: *Cultural Diversity and Water* (2013), *Strategies for Adaptation to Climate Change and its Institutional Effects on Labor Markets* (2013).

Address: Dr. Roberto M. Constantino-Toto, Calz del Hueso 1100, Coyoacan, Villa Quietud, C.P. 04960 Mexico, D.F, Mexico.
Email: roberto.constantino@gmail.com
Website: www.xoc.uam.mx.

Hilda R. Dávila-Ibáñez is a Research Professor at the Economic Production Department at UAM-X. She studied Economics at the Economics Department of the University of Nuevo Leon, holds a Master's Degree in Economics from the New School for Social Research in New York City, and a Ph.D. from the UAM. Her research interests include income distribution, development, growth, and environmental economics focused on the study of water management and policy on which she has published articles and book chapters. She lectures on Macroeconomic for the Master's level in Public Policy and Economics at UAM-X.

Address: Prof. Dr. Hilda R. Dávila-Ibáñez, Calz. del Hueso 1100, Coyoacan, Villa Quietud, C.P. 04960 Mexico, D.F., Mexico.
Email: hdavila@correo.xoc.uam.mx
Website: www.xoc.uam.mx.

About the Authors

Palmira Bueno Hurtado is an agronomist and phytotechnologist from the Agricultural College of the State of Guerrero with an M.Sc. from the Autonomous University of Chapingo. Currently, she works as a researcher at the Integrated Watershed Management INIFAP. Her research focuses on hydrological processes modeling.

Address: Ms. Palmira Bueno Hurtado, CENID RASPA, INIFAP, Km 6.5 Margen Derecha Canal Sacramento Zona Industrial C.P. 35140 Gomez Palacio, Durango, Mexico.
Email: Bueno.palmira@inifap.gob.mx
Website: www.inifap.gob.mx.

Ernesto A. Catalán-Valencia is an Agricultural Engineer in Irrigation from the University of Sonora with an M.Sc. in Hydrosciences from the Postgraduate College and a Ph.D. in Agronomy from New Mexico State University. He is a researcher on water use engineering and an adviser for the research on water use and management in irrigation districts. Since 1984, he has been a researcher for the CENIDs-RASPA-INIFAP in the areas of irrigation engineering, soil physics and agro-meteorology, with multiple publications in journals with national and inter-national circulation, book chapters, and directed theses. Among his research areas are the use of simulation and modeling on subjects such as the design, evaluation, and management of irrigation systems; the flow of water, solutes and energy in the ground; evapotranspiration as well as gas exchange and stomatal control of vege-tated surfaces.

Address: Ernesto A. Catalán-Valencia, CENID RASPA INIFAP Km 6.5 Margen Derecha Canal Sacramento Zona Industrial C.P. 35140 Gomez Palacio, Durango, Mexico.
Email: catalan.ernesto@inifap.gob.mx
Website: www.inifap.gob.mx.

Gerardo Delgado-Ramírez is an Agricultural Engineer in Irrigation from the Agrarian Autonomous University Antonio Narro (UAAAN). His thesis focused on a methodology for the design of irrigation systems with pipe gates. He collaborated

© The Author(s) 2016
R.H. Pérez-Espejo et al. (eds.), *Water, Food and Welfare*,
SpringerBriefs in Environment, Security, Development and Peace 23,
DOI 10.1007/978-3-319-28824-6

with the Integral Transfer Technology Program in Irrigation in the Laguna Region (PROTTIR) implemented by the Agricultural and Livestock Chamber of Torreon and the Farmers Association of the Laguna Region of Durango. He holds a Master's Degree in Agricultural Sciences from UAAAN. His thesis focused on the mathematical analysis of hydrodynamic model of surface irrigation in the Comarca Lagunera. He is a researcher at the INIFAP.

Address: Gerardo Delgado-Ramírez, INIFAP. CENID RASPA INIFAP Km 6.5 Margen Derecha Canal Sacramento Zona Industrial C.P. 35140 Gomez Palacio, Durango, Mexico.
Email: Delgado.gerardo@inifap.gob.mx
Website: www.inifap.gob.mx.

Gerardo Esquivel-Arriaga holds an M.Sc. in Natural Resources and Environment from UACH. He currently works as a researcher in the Integrated Basin Management of the INIFAP. His research area is focused on the modeling of hydrological processes.

Address: Gerardo Esquivel-Arriaga, INIFAP. CENID RASPA INIFAP Km 6.5 Margen Derecha Canal Sacramento Zona Industrial C.P. 35140 Gomez Palacio, Durango, Mexico.
Email: Esquivel.gerardo@inifap.gob.mx
Website: www.inifap.gob.mx.

Jaime Garatuza-Payán is Ph.D. in Hydrology from the University of Arizona in 1999. He has received awards, scholarships, and research grants from the Universities Space Research Association, the UN, and different sources (USRA, ONU, USDA, PROMEP, NASA, CONACYT, NCAR, Sonora State Government, CONAGUA, NOAA, WMO, NSF, UE and Fullbright, European Community). He joined the Sonora Institute of Technology in 1990, where he is currently Senior Fellow C, Director of the Department of Natural Resources, level Division, Level I at the National Research System and member of the Mexican Academy of Sciences. His research interests include mainly those related to hydrometeorology and hydroecology and the use of modeling, remote sensing techniques, and GIS to help assess and manage water resources. In this, his contributions have pioneered in Latin America. He has authored or co-authored more than 50 publications and has supervised the completion of three theses, sixteen Master's degrees and two Ph.D.

Address: Jaime Garatuza-Payán, 5 de Febrero 818 Sur, Col. Centro, C. P. 85000 Ciudad Obregon, Sonora, Mexico.
Email: garatuza1@gmail.com
Website: www.itson.mx.

Eugenio Gómez-Reyes is Research Professor "C" for the Department of Process Engineering and Hydraulics at UAM-I. He holds an M.Sc. and a Ph.D. from the State University of New York. He teaches Engineering Hydrology at UAM-I and has coordinated research projects and directed theses at graduate levels on hydrology, hydrodynamics, and water quality in receiving bodies. He is the author of several articles on water management in the basin of the Valley of Mexico; rainwater harvesting for aquifers recharge; detecting of invisible leaks in distribution networks for drinking water; model of water management and urban development in the Metropolitan Area of Mexico's Valley. He is also the author of numerical models for the simulation of hydrological and hydrodynamic processes in basins and superficial and underground water bodies.

Address: Eugenio Gómez-Reyes, San Rafael Atlixco 186, Vicentina, Iztapalapa, C. P. 09340 Mexico, D.F., Mexico.
Email: egr@xanum.uam.mx
Website: www.izt.uam.mx.

Anne M. Hansen has a Ph.D. from UNAM and is a full-time researcher at IMTA. The Hydrogeochemistry Laboratory was created under her leadership to conduct research on the migration of contaminants and the formulation of preventive, corrective, and conservative water measures. She teaches and leads graduate theses on environmental engineering and earth sciences. She is a member of the National Researchers System currently Level II. She participates in publishing and research resources allocation committees and the Working Group for Environmental Monitoring and Assessment of the Commission for Environmental Cooperation of North America. She participated as alternate IMTA representative for the National Implementation Plan of the Stockholm Convention. She won the Merit Award 2011 of the Council of Science and Technology for the state of Morelos. She is a founding member of the National Institute of Geochemistry.

Address: Anne M. Hansen, Paseo Cuauhnahuac 8532, Col. Progreso, C.P.62550 Jiutepec, Morelos, Mexico.
Email: ahansen@tlaloc.imta.mx
Website: www.imta.gob.mx.

Thalia Hernández-Amezcua holds an undergraduate Economics degree from the University of Guadalajara and Master's in Economics specializing in natural resources and sustainable development from UNAM. She is currently working at the National Institute of Ecology and Climate Change (INECC) as Head of the Department of Economic Instruments. Her work deals with the design and analysis of studies for the identification and implementation of appropriate mitigation and climate change adaptation measures (NAMA) for Mexico. She has worked in several consulting projects for the National Water Commission, the Mexican Institute of Water Technology, the World Meteorological Organization and the World Bank.

Address: Thalia Hernández-Amezcua, Periferico 5000, Col. Insurgentes Cuicuilco, Delegacion Coyoacan, C.P. 04530, Mexico, D.F., Mexico.
Email: thalia.hernandez@inecc.gob.mx
Website: www.inecc.gob.mx.

Delia Montero is a full-time Research Professor "C" for the Economics Department at UAM-I. She is member of the National Researchers System Level I. She holds a Ph.D. in Economics from the School of Higher Studies in Latin America (IHEAL) at the Sorbonne Nouvelle Paris III, France. She holds a Master's degree in Rural Development at the Institute for Economic and Social Studies (IEDES) in Paris, France and a BA in International Relations at UNAM. She was a member of the Divisional Board of Social Sciences and Humanities and of the Adjudication Board of the International Council for Canadian Studies (ICCS). She earned a scholarship from the Government of Quebec to do research and was a visiting professor at the Department of Political Sciences at the University of Quebec at Montreal (UQAM). She is a founding member of the Water Forum of UAM. She currently coordinates the Quality and Water Reduction Demand Project in Mexico's Federal District (DF) funded by the DF's Institute of Science and Technology.

Address: Delia Montero, San Rafael Atlixco 186, Vicentina, Iztapalapa, C. P. 09340 Mexico, D.F., Mexico.
Email: del@xanum.uam.mx
Website: www.izt.uam.mx.

Úrsula Oswald-Spring is a researcher at CRIM-UNAM and first Chair on Social Vulnerability at the United Nations University. She studied medicine, psychology, philosophy, languages, anthropology, and ecology in Madagascar, Paris, Zurich, and Mexico. She holds a Ph.D. in Social Anthropology and Ecology from Zurich University. She is a member of the IPCC, the World Social Science Report and a reporting GEO-5-UNDP. She was coordinator for the Scientific Network on Water of CONACYT and the project: "Climate Change and Integrated Management of River Basins Yautepec". She was Minister of the Environment for the State of Morelos (1994–1998) and the first Attorney of Ecology (1992–1994), President of the International Association of Peace (1998–2000) and General Secretary of the Latin American Council of Peace Research (2002–2006). She has published 47 books, 318 articles and chapters in scientific books. Awards include Sor Juana Ines de la Cruz (1990), Fourth Decade of UN Development, Academic Women (UNAM 1991), and Women of the Year 2000.

Address: Úrsula Oswald-Spring, Av. Universidad s/n, Circuito 2, C.P. 62210, Cuernavaca, Morelos, Mexico.
Email: uoswald@gmail.com
Website: www.crim.unam.mx.

Francisco Javier Peña-De Paz has a Ph.D. in Social Sciences specializing in Anthropology. He is a research fellow for the Water and Society Program at Colegio de San Luis and a member of the International Water Justice Alliance. His areas of study include: risk, cultural diversity and territorial construction, water crisis and water justice. He has conducted research in different parts of Mexico. He is a collaborator of the Integrated Ecosystem Management Program, Advisor of the Board of the Altiplano Basin, and Member of the National Researchers System. Recent publications include: "Indigenous Peoples and Water Resource Management in Mexico" (2006), "Vocations and risks of a Disputed Territory. Actors, representations and Arguments Against the Minera San Xavier" (2008, co-author); "Indigenous Peoples and Local Water Conflicts" in *El agua en Mexico*: *cauces y rios* for the Mexican Academy of Sciences (2010); "Threats to a Sustainable Future: Water Accumulation and Conflict in Latin America" (2011, co-author). He is a Member of the Scientific Network on Water for the CONACYT.

Address: Francisco Javier Peña-De Paz, Parque de Macul No. 155, Fracc. Colinas del Parque, C.P. 78299 San Luis Potosi, S.L.P., Mexico.
Email: desiertosol2@gmail.com
Website: www.colsan.edu.mx.

Patricia Phumpiu-Chang holds a Ph.D. on Water and Soil Resources and Expertise in Environmental Engineering from the Royal Institute of Technology (KTH), Sweden. She holds a Master's degree in Environmental Policy and regulation from the London School of Economics and Political Science (LSE), UK, and in Urban Planning from the University of Pennsylvania, USA. She is a professor at ITESM (Mexico) and researcher at the Water Center for Latin America and the Caribbean. She is Coordinator for the research group on Environmental Policy Analysis and Water Policy, governance issues in water management, institutional analysis, regulation and role of collective participation in water governance, and is lead coordinator for the Water, Sanitation and Hygiene Project for Community Management taking place throughout three states in Mexico. Her professional experience includes consulting for government agencies in South and Central America, Europe, and Asia.

Address: Patricia Phumpiu-Chang, Agatan, Agatan 22, Sundbyberg, Stockholm, 17262, Sweden.
Email: patricia.phumpiu@itesm.mx
Website: www.itesm.mx.

Abel Román-López is an Industrial Mechanical Engineer from the Technological Institute La Laguna # 13 and Master in Irrigation and Drainage from the Agrarian Autonomous University Antonio Narro. He has been an Irrigation Engineering Researcher at the CENIDs-BARK-INIFAP since 1980. He has published in national and international journals, directed theses, technical and scientific brochures, and a technical book, and the generation of computer software for water efficiency and energy in the pressurized irrigation systems and deep wells of underground mining. He has delivered training courses and led advanced studies on pressurized irrigation technology, evaluation and management of pressurized irrigation systems with alternative energy and renewable resources.

Address: Abel Román-López CENID RASPA INIFAP Km 6.5 Margen Derecha Canal Sacramento Zona Industrial C.P. 35140 Gomez Palacio, Durango, Mexico.
Email: roman.abel@inifap.gob.mx
Website: www.inifap.gob.mx.

Ignacio Sánchez-Cohen is an Agronomist Engineer from the University of Sonora and Ph.D. in Watershed Management on Physical Aspects of Arid Land from the University of Arizona. He is member of the National Center for Disciplinary Research in Foreign Soil Water Atmosphere Plant (CENIDRASPA) of the National Institute of Livestock Agriculture and Forestry Research (INIFAP). He is a researcher Level II for the National System of Researchers and national leader for the Integrated Watershed Management Research Program (INIFAP) and has been a member of the Board of Governors for that Institute. His research interests include integrated water management, hydrological modeling of processes and systems for decision-making. He is a member of the Academic Committee of the Water Theme Network of the National Council of Science and Technology. He is the author of several national and international journal articles indexed in the International Scientific Index, six books and has co-authored four publications. He was Director of CENIDRASPA from 1997 to 2005.

Address: Ignacio Sánchez-Cohen CENID RASPA INIFAP Km 6.5 Margen Derecha Canal Sacramento Zona Industrial C.P. 35140 Gomez Palacio, Durango, Mexico.
Email: sanchez.ignacio@inifap.gob.mx
Website: www.inifap.gob.mx.

Germán Santacruz-de León is a professor and researcher for the "Water and Society" program at the Colegio de San Luis and is a member of the National System of Researchers. He holds a Ph.D. in Environmental Sciences from UASLP and Master's degrees in Hydraulic Engineering from UNAM as well as Environmental Engineering from the National Polytechnic Institute. He is an Agronomist Engineer specializing in Irrigation from the Chapingo Autonomous University. His research focuses on the social and environmental problems associated with the use, handling and management of water resources. His publications include "The Mirage of the Watershed as an Area for the Management of Surface Water Resources regarding the Valles River, Huasteca. Mexico". Research Collection, El Colegio de San Luis, A.C.; "Environmental problems and social conflicts over water use in the basin of the Suchiate River" and "Government lethargy in the management of water resources: the case of the Suchiate River basin" in *Between Springs and Unleashed Rivers: Paradoxes of Border Hydro Policies* (Mexico-Guatemala). Kauffer Michel, Edith (coord.)

Address: Germán Santacruz-de León, Parque de Macul No. 155, Fracc. Colinas del Parque, C.P. 78299 San Luis Potosi, S.L.P., Mexico.
Email: gsantacruz@colsan.edu.mx
Website: www.colsan.edu.mx.

Andrea Santos-Baca graduated from the Economics Department at UNAM and holds a Master's degree in Social Sciences from the Latin American Faculty of Social Sciences, Mexico campus. She has worked on consumer issues and economic dependence and recently on the effect of free trade on the dietary patterns in Mexico. In March 2012, she participated in the "Collaborative Workshop: Poverty and Peasant Persistence in the Contemporary World" organized by CROP/El Colegio de Mexico/UAM. In February 2012 she was a member of the organizing committee for the 7th International Seminar on Animal Reproduction and Production of Milk and Meat: Artisan Food Production and Marketing, Siraplec-UAM Xochimilco and in 2007–2009 wrote the chapters on water use, pesticides, and proposals of agro-environmental policy in the book *Agriculture and Water Pollution*, coordinated by Rosario Pérez-Espejo, Ph.D.

Address: Andrea Santos-Baca, Calle Balboa 608, torre A, Depto. 401., Col. Portales, Delegación Benito Juarez, México DF, CP 03300, Mexico.
Email: hibridos@hotmail.com.

About this Book

This book is a result of the cooperation of 16 researchers from 11 Mexican academic institutions. It addresses the following topics: the contemporary model for water management and alternative approaches; socioeconomic framework, water policy and institutions; water use for food purposes, water resources inventory and irrigation; manifestations of welfare loss, water prices; change in the dietary pattern and water security; hydrological stress and pressures on water availability; problems of groundwater management; vulnerability and climate change; water demand of major crops; gray water footprint and water pollution; gray water footprint and mining; virtual water and food trade; estimates of the water footprint of four key cereals, forages, livestock, and bottled drinks and beverages.

- Own estimates of the water footprint and a critical application of this methodology;
- An overview of water management, water policies, and institutions;
- An analysis of the major factors affecting water resources in Mexico.

Contents:

Part I: Linking Water Management, Food Policy and Welfare—*R.M. Constantino-Toto*: Contemporary Model for Water Management and Alternative Approaches—*H.R. Dávila-Ibáñez; R.H. Pérez-Espejo; T. Hernández-Amezcua*: Socioeconomic Framework—*R.H. Pérez-Espejo; T. Hernández-Amezcua; H.R. Dávila-Ibáñez*: Water Policy and Institutions—*R. Pérez-Espejo; T. Hernández-Amezcua; H.R. Dávila-Ibáñez*: Water Use for Food Purposes—*E. Gómez-Reyes—J. Oswald-Spring Garatuza-Payán—R.M. Constantino-Toto*: Water Resources Inventory and Implications of Irrigation Modernization—*Ú. Oswald-Spring*: Manifestations of Welfare Loss—*R.M. Constantino-Toto*: Prices and Water. A Strategy with Limited Effectiveness.

Part II: Pressures on Water Availability, Its Use and Welfare Effects—*E. Gómez-Reyes*: Water Use Pattern—*A. Santos-Baca*: Change in the Dietary Pattern and Water Security—*P. Phumpiu-Chang*: Hydrological Stress and Pressures on Water Availability—*E. Gomez-Reyes*: Problems Associated with Groundwater Management—*H.R. Dávila-Ibáñez; R.M. Constantino-Toto*: Vulnerability and Climate Change.

© The Author(s) 2016
R.H. Pérez-Espejo et al. (eds.), *Water, Food and Welfare*,
SpringerBriefs in Environment, Security, Development and Peace 23,
DOI 10.1007/978-3-319-28824-6

Part III: Methodology for Analyzing Water Footprint and Virtual Water—*I. Sánchez-Cohen; E. Catalán-Valencia; J. Garatuza-Payán*: Water Demand of Major Crops. A Methodology—*A.M. Hansen*: Gray Water Footprint and Water Pollution —*G. Santacruz-De León; F.J. Peña-de Paz*: Gray Footprint and Mining: Impact of Metal Extraction on Water—*T. Hernández-Amezcua; A. Santos-Baca*: Considerations on Virtual Water and Agri-food Trade.

Part IV: Applying the WF Approach for Impact Analysis on Sectors and Regions —*R. Pérez-Espejo; T. Hernández-Amezcua*: Water Footprint of Four Cereals in Irrigation District 011—*I. Sánchez-Cohen; G. Delgado-Ramírez; G. Esquivel-Arriaga; P. Bueno-Hurtado; A. Román-López*: Forage Water Footprint in the Comarca Lagunera—*R.H. Pérez-Espejo; T. Hernández-Amezcua*: Water Footprint in Livestock—*R.M. Constantino-Toto; D. Montero*: Water Footprint of Bottled Drinks and Food Security.

More on this book is available at: http://www.afes-press-books.de/html/ SpringerBriefs_ESDP_23.htm.